Earth: Inside and Out

Earth: Inside and Out

Edmond A. Mathez, Editor

AN AMERICAN MUSEUM OF NATURAL HISTORY BOOK

The New Press, New York

Cover: Chris McKay looking down a drill hole cut into the frozen surface of an Antarctic lake.

Right: Earth as seen from space.

Published in the United States by The New Press, New York, 2001
Distributed by W. W. Norton & Company, Inc., New York

LIBRARY OF CONGRESS CATALOGING-IN-PUBLICATION DATA
Earth: Inside and Out/Edmond A. Mathez, editor.
p. cm.
ISBN 1-56584-595-1 (pbk.)
1. Geology. 2. Earth. I. Mathez, Edmond A.
QE26.2.E274 2001
550—dc2100–136454

The New Press was established in 1990 as a not-for-profit alternative
to the large, commercial publishing houses currently dominating
the book publishing industry. The New Press operates in the public
interest rather than for private gain, and is committed to publishing,
in innovative ways, works of educational, cultural, and community
value that are often deemed insufficiently profitable.

The New Press, New York
450 West 41st Street, 6th floor
New York, NY 10036

www.thenewpress.com

Printed in England

9 8 7 6 5 4 3 2 1

Contents

9   **Foreword:** Ellen V. Futter

10 **Acknowledgments**

12 **Preface: Earth: Inside and Out** Edmond A. Mathez

Section One: How Has the Earth Evolved?

16 **Introduction** Edmond A. Mathez

20 **An Earth Moon Mystery** Robert A. Fogel

26 **Origin and Evolution of the Continents** Roberta L. Rudnick

32 **Life and the Evolution of Earth's Atmosphere** Stephen J. Mojzsis

41 **Case Study: Retrieving a Stromatolite from the Sahara Desert**

45 **Case Study: Zircon Chronology:
       Dating the Oldest Material on Earth**

49 **Profile: Arthur Holmes: Harnessing the Mechanics of Mantle
       Convection to the Theory of Continental Drift**

Section Two: How Do Scientists Explore the Inner Earth?

52 **Introduction** Edmond A. Mathez

56 **Global Seismic Tomography:
       A Snapshot of Convection in the Earth** Robert D. van der Hilst

62 **Convection in the Core and the Generation
       of the Earth's Magnetic Field** Gary A. Glatzmaier

68 **Mantle Convection** Peter Bunge

73 **Case Study: Ultra-High-Pressure Experimentalist
       Who Studies the Deep Earth**

77 **Profile: Inge Lehmann: Discoverer of the Earth's Inner Core**

Section Three: Plate Tectonics in Action

80   **Introduction** Edmond A. Mathez

84   **The Structure of Mountain Ranges** Peter H. Molnar
90   **Averting Earthquake Surprises in the Pacific Northwest** Brian F. Atwater
94   **Hawaii and Hotspots: A Window to the Deep Mantle** Steven L. Goldstein
100 **The Hazards of Volcanoes** Haraldur Sigurdsson
105 Case Study: **Forecasting Earthquakes Using Paleoseismology**
109 Case Study: **A Conversation with Jacques Malavieille**
113 Profile: **Harry Hess: One of the Discoverers of Seafloor Spreading**

Section Four: How Do Scientists Read the Rocks?

116 **Introduction** Edmond A. Mathez

118 **The Erosion of the Grand Canyon** George Billingsley
124 **Death Among the Dunes:**
        A Dinosaur Murder Mystery Lowell Dingus and David B. Loope
129 Case Study: **Mapping Mt. Rainier**
133 Profile: **James Hutton: The Founder of Modern Geology**

Section Five: What Causes Climate and Climate Change?

136 **Introduction** Edmond A. Mathez

140 **The Oceans' Role in Climate** Martin H. Visbeck
146 **Predicting El Niño** Mark A. Cane
154 **Global Warming** Charles F. Keller
160 **An Ice Core Time Machine** Paul A. Mayewski
169 **Case Study: Studying Tree Rings to Learn About Global Climate**
173 **Profile: Milutin Milankovitch: Seeking the Cause of the Ice Ages**

Section Six: Why Is the Earth Habitable?

176 **Introduction** Edmond A. Mathez

178 **Earth: The Goldilocks Planet** Rachel Oxburgh
184 **Black Smokers: Incubators on the Seafloor** Deborah S. Kelley
190 **Resources of the Earth:**
         Where Do Metals Come From? James Webster
197 **Case Study: Looking for Life in Antarctica... and on Mars**
201 **Case Study: Mapping Hot Springs on the Deep Ocean Floor**
205 **Profile: Harold C. Urey: Discoverer of Deuterium and**
         Investigator of the Origin of Life, the Origin of the
         Planets, and the Climate of the Early Earth

207 **About the Gottesman Hall of Planet Earth**
214 **Resources**
217 **About the American Museum of Natural History**
218 **Contributors**
223 **Credits**
226 **Glossary**
237 **Questions**

Foreword   Ellen V. Futter

Since its founding in 1869, the American Museum of Natural History has put the world on display. The twin pillars of our mission have always been science and education. Today, the Museum is one of the world's leading research institutions in the natural sciences. Over the years, our scientists—now a staff of more than 200 men and women—have gone on more than 100 expeditions a year. They collect evidence from all over the globe in their effort to answer questions about such fundamental scientific and human issues as the origins of the universe, Earth and life, who we are and where we fit. In addition to research, our scientists have a related responsibility—to interpret science for the general public. The exhibitions at the Museum have been conceived and curated by scientists who are committed to putting the evidence—"the real stuff" in front of the public.

As we move into the twenty-first century, we at the Museum are filled with a renewed dedication to our mission. To many people, science today seems too removed and too difficult to understand, yet the need has never been greater for a public that is well informed about science. Through its educational initiatives and exhibitions, the Museum seeks to narrow the gap between what people know and what they need to know about science. To that end, in 1997 the Museum launched the National Center for Science Literacy, Education, and Technology to extend the Museum's resources beyond its walls to a national audience.

Science is exploration. Scientists work at the frontier—at the border of the known and the unknown. This book series, through the words of working scientists, enables non-scientists to share the excitement of cutting-edge science— the excitement of discovery. The series includes four volumes that expand the themes covered in many of our major new exhibitions. Our exhibitions always embody a scientist's vision and point of view. In the same way, each book in this series is "curated"—researched, organized, and introduced—by one of the Museum's scientists. Each book features a selection of essays written by leading scientists who have made significant contributions to the field. The essays are supported by case studies and profiles of important people and events.

This volume, *Earth: Inside and Out*, focuses on the Earth and the phenomena that shape our planet to expand the themes addressed in our new Gottesman Hall of Planet Earth. Other volumes in this series include: *Epidemic! The World of Infectious Disease,* which explores the themes of the major special exhibition at the Museum in 1999 by the same name; *The Biodiversity Crisis: Losing What Counts,* which focuses on biodiversity, and the interrelatedness of all living things on our planet to expand the themes addressed in the Hall of Biodiversity; and *Cosmic Horizons: Astronomy at the Cutting Edge*, which explores the mysteries and wonders of the universe to expand the themes addressed in our new Cullman Hall of the Universe.

This series illustrates our continuing commitment to connect the general public with the natural world. We cannot send real specimens to every home and classroom, but we can bring the ideas, concerns, and questions of working scientists directly to you. We hope these books provide a valuable resource that will prepare tomorrow's leaders to make informed decisions about the world we all share.

Ellen V. Futter is President of the American Museum of Natural History.

Acknowledgments

This book was produced by the National Center for Science Literacy, Education, and Technology, American Museum of Natural History.

Ellen V. Futter, President

Myles Gordon, Vice-President of Education

Nancy Hechinger, Director of the National Center for Science Literacy, Education, and Technology

The National Center would like to acknowledge the National Aeronautics and Space Administration for its general programmatic support and for the support of this book series.

Editorial Staff:
Science Editor: Rosamond Kinzler
Associate Science Editor: Heather Sloan
Writer: Ashton Applewhite
Education Editor: Michele Williams

Production Staff:
Project Director: Caroline Nobel
Creative Director: Patricia Abt
Producer: Allison Alltucker
Production Managers:
Ellen Przybyla and Tom Baione
Production Coordinator: Eric Hamilton
Production Assistant: Ethan Davidson

Design by Sheena Calvert, parlour design, NY

We would like to thank Diane Wachtell, Ellen Reeves, Barbara Chuang, Leda Scheintaub, Gary Tooth, and Fran Forte at The New Press.

Exterior of the American Museum of Natural History.

The Science of the Earth Is Fundamentally Historical

View of the Grand Canyon in Arizona.

Preface: Earth: Inside and Out

Edmond A. Mathez

Unlike most other sciences, the science of the Earth is fundamentally historical. Geologists look at rocks in order to deduce geological history. But that history is more than the story of how the Earth has evolved. It is also a way to understand the processes that have shaped the Earth. A historical approach is necessary because these processes generally occur across distances and lengths of time that are far beyond human experience. For example, the complex circulation patterns of the oceans and atmosphere that cause climate variations are far too vast to be reproduced in a test tube, or a lab, or even an airplane hangar. Instead, we need to observe the whole planet. Even then, although climate change occurs at lightning speed compared to most global geologic processes, some circulation patterns are too slow or too irregular to be seen over the course of a lifetime. So, to understand how the climate system works, we must study the records of past climate.

Although this geologic record is incredibly ancient, it has only come under intense scrutiny relatively recently. Compared to other natural sciences, earth science is young; it emerged as a true science at the end of the eighteenth century. This is not to say that natural philosophers, as most scientists were referred to at the time, didn't look at rocks and wonder how they had formed. But, just as it is hard to grasp what a rock can tell you until you've had a few lessons in earth science, these people had no idea how to read a rock's message.

Part of the reason for this, at least in Christian cultures, was the deeply-rooted concept that the Earth was young. In the seventeenth century, a detailed and scholarly investigation of the Bible by James Ussher, then Archbishop and Primate of Ireland, yielded a precise age. According to Ussher, the Earth formed on 4,004 B.C. (at the "entrance of night" on October 22, to be exact), thus making it about 5,700 years old. (Although by no means the first or only Biblical age, Ussher's was to acquire an air of official sanction because it appeared as an annotation in many of the King James translations of the Bible.) This led to the conclusion that all the Earth's surface features, including sedimentary rocks containing obvious forms of life in the form of fossil shells, were a consequence of the Biblical flood. There the matter rested for many years. Near the end of the eighteenth century, however, natural philosophers—including Georges Cuvier in France and James Hutton in Scotland—began to doubt that the Earth could be so young, but for different reasons. As observations of rocks and fossils accumulated, they made less and less sense in the Biblical context.

James Hutton, who became known as the father of geology, was probably most influential in giving shape to this young science. His observations, which implied that the Earth must be very old, marked the beginning of the systematic science. Its first flourishing extended through most of the nineteenth century, thanks in large part to Charles Lyell, who embraced Hutton's ideas and wrote the first and highly influential textbook, *Principles of Geology*, published between 1830 and 1833 in three volumes, which went through twelve editions in Lyell's lifetime. For the first time, geologists took systematic note of the rocks around them. They began to apply the principles of the new science, such as the Principle of Superposition, which states that younger rocks lie on top of older ones, and the Uniformitarian Principle, which states that the processes that occur

Edmond A. Mathez is a Curator in the Department of Earth and Planetary Sciences at the American Museum of Natural History.

around us today are those that occurred in the past. These may seem obvious, and indeed they are simple matters of common sense. But they led to profound notions, such as the idea that the Earth is ancient. The natural philosophers began to realize there were answers to questions such as how mountains form and where lava comes from. They began to wonder what lay beneath the oceans and deep under foot. In evidence such as the debris scattered by the glaciers that had once covered much of Europe, they saw that the Earth had undergone massive changes. Quite naturally, they sought to write the history of the planet from this new perspective. The nineteenth century was the "golden age" of this nascent science because its investigations established the foundation of modern geology.

A remarkable "revolution" in the science of the Earth began about 1960 (and may still be going on) when the concept of plate tectonics emerged. This theory states that the Earth's surface is made up of rigid plates that move, thus accounting for the existence of ocean basins, continents, mountain belts, volcanoes, and earthquakes. In short, it exerts a fundamental control on the shape of this planet. The concept was revolutionary because it provided the context for phenomena that until then had not seemed related. Plate tectonics now hardly qualifies as a theory, since satellite measurements have proven that distant regions of the Earth are moving relative to each other.

The revolution, however, did not end with the discovery of plate tectonics. Shortly afterward, we humans visited the Moon, and this inspired us to think anew about the early history of the solar system, how our planet formed, and what the early Earth was like. Then, like all other

sciences, earth science exploited the new power of computers. They guided satellites, but more importantly they enabled satellites to look back on the Earth from space. That is, computers allowed us to send, store, and otherwise handle the digital data that make up the "photographs" and other observations collected by satellites, as well as data from instruments such as seismographs that monitor activities in the Earth's interior. For the first time, we began to "see" Earth processes at work on a global scale. For example, measurements of the state of the sea surface yielded new insights into the phenomenon of El Niño. Perhaps one way to characterize the new science is that we have come to understand how to place events and processes in a global context. This means that we are beginning to understand how the Earth works.

To those of us in this new field of earth science, it seems as if every day brings a new and interesting finding that was out of our grasp only a short time before. The scientists whose essays appear in this book are among the new discoverers. If they are like me, they find the process of discovery as exciting as the discoveries themselves.

The essays in this book are loosely organized around the five questions posed in the American Museum of Natural History's Gottesman Hall of Planet Earth: How has the Earth evolved? How do scientists read the rocks? Why are there ocean basins, mountains, and continents? What causes climate and climate change? Why is the Earth habitable? This collection is far from a comprehensive study of the Earth. The science is just too vast, and there are surely other topics as important as those we have chosen. The essays do, however, paint a broad portrait

of a science at an exciting moment in its evolution. If nothing else, I hope they will give you a sense of how geologists think about the Earth.

Who cares about earth science, and why is it relevant to you? Well, I have two answers. The first one is easy. Humanity has acquired the power to modify our planet in ways that are profound—and potentially damaging. The state of the Earth affects the everyday lives of many people, in ways that range from the mundane to the catastrophic. Hundreds of thousands of people have died almost instantly in a single earthquake, and a hurricane can wreck thousands of homes. A better understanding of short-term phenomena such as tsunamis, floods, and hurricanes, as well as of longer-term events such as global warming, will greatly benefit millions of people.

These important matters are not what make earth science relevant to me, nor, I suspect, to most of the contributors to this book. Our studies are intensely personal quests into the unknown. It is the ability to use our knowledge and imagination to make new knowledge that gives us an identity and defines us as humans. The research is our art. I hope this book helps you to find yours. ❻

This wall of dark rock on the beach at Siccar Point, Scotland, got its shape when molten rock intruded into a crack in another rock deep under ground. Erosion of the overlying rocks then exposed this slice of rock known as a dike.

Section One: **How Has the Earth Evolved?**

Detail of Banded Iron Formation from
Strathay Township, Ontario, Canada.

Introduction

The processes that have governed the evolution of the Earth generally act over scales of time and space of such vastness that they are difficult to understand. For example, the 10 million years or so that it may have taken to build a mountain chain or move a continent is way beyond our everyday experience. Fortunately, however, 10 million years is not beyond our imagination. In our attempts to decipher the history of our planet, we geologists have had to learn to imagine such vast scales, and to think and talk in terms of them. In this first section, we ask you to imagine along with us as we discuss three elements of this history of the Earth that are extremely large, although in different ways. They are: the origin of the Moon, which is connected to the origin of the Earth; the formation of continents, which do not exist on any of the other planets; and the evolution of the atmosphere, which has been strongly influenced by the emergence of life and has in turn influenced how life evolved. These stories have been chosen for their great interest to earth scientists.

The origin of the Moon also concerns the Earth because, along with the Sun and the other planets and probably their moons, they both formed from the original solar nebula. The nebula was a rotating disk of gas and dust. Within this disk the dust agglomerated into progressively larger particles with ever-greater mass and gravity. In this manner, the Sun and planets that we see today were eventually to emerge. Since humans have set foot on the Moon, one might imagine that the detailed study of the rocks brought back from the Moon tells us exactly how it formed. But according to Robert A. Fogel's account, this is hardly the case.

Also important to the Earth's early evolution was its differentiation into a layered body. In other words, in the early Earth heavy materials sank to the center, and light materials rose to the top. This resulted in the gross structure of the Earth we see today. At its center is a solid inner core, which is surrounded by a liquid outer core. Both are made of dense metal consisting mainly of the elements iron and nickel. Surrounding the outer core is the mantle, which is made up of dense silicate rocks, and above the mantle lies the crust, which is composed of lighter silicate rocks.

In the second essay, Roberta L. Rudnick writes on the origin and evolution of the continents. Continents are particularly important to geologists because they hold practically the entire historical record of the planet. Moreover, how the continents form and how long the process first took are questions central to understanding the Earth's evolution. By radiometric dating of the rocks of the continental crust and by mapping their distributions, we know that most of the continental crust evolved early and rapidly in the Earth's history. To see how the crust has continued to evolve through time, we look at the processes that operate today in mountain belts (see the description of Jacques Malavieille's experiments on building models of mountains in Section Three) and subduction zones, where the Earth's oceanic crust is being subducted, or thrust back into the mantle, and where new continental crust is being made. These zones are marked by volcanic island arcs such as the Aleutians, and mountain belts such as the Andes. Continental evolution is an extremely complicated process, and scientists are still working to fully understand it.

There are a number of ways to determine the precise ages of rocks, including radiometric dating. One of the most fascinating techniques involves studying zircons—a common mineral but one that exists in only miniscule amounts in most rocks. Once formed, zircons are so hard they are almost impossible to destroy. Consequently—although it may be hard to believe—a single zircon crystal, originally formed in the deep core of a mountain, may survive and even record several episodes of erosion, sedimentation, and mountain formation, cycles which can take hundreds of millions of years. The case study about the work of Darrel Henry and Paul Mueller describes how scientists collaborate together to deduce the histories of some of Earth's oldest rocks.

The evolution of the Earth obviously requires some appreciation of geologic time. One of the most important figures in the development of a quantitative geological time scale was Arthur Holmes, the subject of this section's profile. His extensive work in the radiometric dating of rocks and minerals, accomplished through the first half of the twentieth century, is a basis for the modern time scale.

In addition to the rock layers, we should not forget another layer that also originated by differentiation of the early Earth, namely the ocean-atmosphere. Very old rocks suggest that the ocean and atmosphere of the early Earth were much different than our present ones. In particular, they lacked free oxygen. By 3.5 billion years ago, and perhaps even earlier, organisms had evolved that used the Sun's energy to transform carbon dioxide, then in high concentrations in the atmosphere, into the material of the organism itself, giving off oxygen in the process. In other words, these organisms were engaging in photosynthesis. Initially, all the oxygen produced by photosynthesis was absorbed by the ocean, where it caused the iron dissolved in the water to precipitate out as iron oxide, thus creating iron-rich sediments known as banded iron formations. Eventually,

about 1.7 billion years ago, the oceans ran out of iron. At that point, the level of oxygen in the atmosphere was able to increase because there was no more iron left in the ocean to react with it. Scientists have surmised this because banded iron formations are found only in rocks that are more than 1.7 billion years old, while younger sedimentary rocks exhibit characteristics of having formed in an oxygen-rich atmosphere. Although we seldom think of it in this way, the oxygenation of the atmosphere represents a profound way in which life has affected the evolution of the Earth itself. The story of the rise of early life and evolution of the atmosphere is told in the essay by Stephen J. Mojzsis.

The creatures responsible for producing the oxygen were nothing more than tiny microbes. These were the only form of life on Earth until about 1 to 1.2 billion years ago, when multicellular life first appeared. Fortunately for us, the microbes left fossil traces in the form of stromatolites. These can perhaps be described as slimy masses that trapped sediment as they grew. Fossil stromatolites are fairly common in certain areas of old rocks. One such area is in the Western Sahara desert of Mauritania. The case study about Heather Sloan, the Museum scientist in charge of our sample-seeking expedition there, recounts some of her experiences. ❻

To explore Earth's evolution I pose the following questions:

How did the Moon form?

Robert A. Fogel, a Research Scientist in the Department of Earth and Planetary Sciences at the American Museum of Natural History, explores research approaches, and four possible models that could explain the origin of the Moon.

What Earth processes shaped the continents?

Roberta L. Rudnick, a Professor in the Department of Geology at the University of Maryland, describes the formation of continental crust at subduction zones and hotspots and how it evolves.

How has life influenced the evolution of Earth's atmosphere?

Stephen J. Mojzsis, an Assistant Professor in the Department of Geological Sciences at the University of Colorado, Boulder, explains how evolving life has changed the chemistry of the atmosphere.

A close-up view of an outcrop of fossilized stromatolite from Mauritania, Africa.

An Earth Moon Mystery

Robert A. Fogel

Detective Columbo awoke in the middle of the night to the gravelly voice of his Chief on the other end of his telephone; he knew there would be no sleep for the rest of the night. He drove swiftly to 238 Uranium Lane and pushed his way through the bystanders milling around behind the police line. The questions on their minds were all the same: Who was the murderer, how did he do it, and why?

A small army of experts on everything from ballistics to forensics had already descended on the scene, searching the entire area

Robert A. Fogel is a Research Scientist in the Department of Earth and Planetary Sciences at the American Museum of Natural History.

The near side of the moon.

with a fine-tooth comb for the smallest clue. Pictures were snapped from every possible angle and samples of all materials even remotely relevant to the crime were taken and catalogued for future analysis. At 6 AM an exhausted Columbo made his way to a local diner and sipped the first coffee of the day. All this nosing around had left him even more stumped over this homicide than when he had first arrived on the scene a few hours earlier.

About twenty miles away, another early-morning coffee was being poured. Dr. Thera Noom, a scientist at the Museum of Natural History, had also been up all night, and like Columbo, she had a mystery on her hands. A moon had been placed around her planet and she just had to know how it had gotten there. By contrast, her problem made Lieutenant Columbo's look like a piece of cake. Over several decades Dr. Noom and her colleagues had collected enough scientific evidence to dwarf that of even the most complicated murder cases. True, the possibilities had narrowed over the years, just as in any good murder investigation, but the ones that remained still left much room for doubt. She poured some more coffee and got back to work.

Dr. Noom lives on a planet around a main sequence star in a far-flung arm of one of the many billions of spiral galaxies in the universe. The spiral galaxy is called the Milky Way, the star her planet orbits is called the Sun, and the name of her planet is Earth. She is one of thousands of planetary scientists around the globe trying to solve the same mystery: how did the Moon form? This question has long been asked by almost every society and culture and a vast number of myths, legends, and scientific theories have been formulated to address it.

Although their problems are different, Dr. Noom and Detective Columbo have more in common than one would suspect. Both collect as much

evidence—or data—as possible, come to conclusions about small pieces of their puzzle, go out and collect more data, and so on. Sometimes just reordering the evidence in a specific logical sequence brings new insights, but it's usually back to the lab or scene of the crime for another piece of data. For both detective and scientist, solving small pieces of the puzzle ultimately leads to the entire solution. Both require enough small pieces to make the entire picture become clear.

Lieutenant Columbo's evidence takes the form of objects, images, and verbal accounts relating to the homicide. His files pile up with descriptions of bullets, chemical analyses of fingernail scrapings, photographs of the body, and so on. Dr. Noom's evidence is very different: it consists of the elements of the periodic chart. Her data describes how these elements are put together within various compounds—called minerals—that are found in the rocks and soils from the Earth and the Moon. Fortunately, Moon rocks brought back to the Earth by six Apollo missions are available for her analyses.

It's Columbo's job to come up with the most plausible set of circumstances that led to the murder. These circumstances must be based on as much evidence as possible. Likewise, Dr. Noom's job is to theorize about the most plausible set of physical and chemical events that led to the Moon's orbit around the Earth. Her theories, too, must be based on as much data as possible.

One big difference makes Dr. Noom's job more complicated than that of any homicide detective: the events that led to the formation of the Moon occurred 4.5 billion years before Columbo's homicide—give or take 50 million years. A lot can happen in 4,500,000,000 years to obscure evidence of even the most violent of physical events. Working to Dr.

Noom's advantage is the fact that the Moon has been a largely dead world, geologically speaking, for the last three-and-a-quarter billion years. On Earth, by comparison, plate tectonics and its tumultuous combination of volcanism and earthquakes, plate spreading and subduction, continuously recycle and reshape the surface of the planet.

Yet the first 1.25 billion years of lunar history were a time of violent transformation, and it's the events that occurred during this period that obscure a clear reading of the Moon's formation. Soon after its birth, the Moon went through a period of massive melting that encompassed the entire Moon down to a depth of several hundreds miles. This sea of molten rock was hot indeed, with temperatures starting at 1,200°C (2,192°F) and rising. This sea of hot lunar lava is called the Lunar Magma Ocean, and its existence has placed an indelible stamp on the Moon's appearance.

You see the consequences of the Lunar Magma Ocean every time you look at the Moon. From the Earth, the Moon looks like a white circle with several large dark areas in the interior. The bright whitish-gray areas are mountains called the Lunar Highlands, and the flatter, grayish-black areas are called the Lunar Mare (pronounced mah-ray). The Lunar Highlands are mostly composed of a rock type called anorthosite, after the mineral anorthite that makes up a large part of it. These anorthosites date from the first few hundred million years of lunar history. As the Lunar Magma Ocean cooled, minerals crystallized inside it. Those minerals that were denser than the surrounding magma sank by the force of gravity to the bottom of the Magma Ocean. Likewise, those minerals that were less dense than the surrounding magma rose to the top of the Lunar Magma Ocean. Anorthite was the major

light mineral that floated in this way, and its accumulation at the top of the Magma Ocean formed the anorthosites of the Lunar Highlands. The whitish color of anorthite contributes to the whitish-gray appearance of the Lunar Highlands. The contents of the Lunar Mare were deposited much later, between 4.2 and 3.2 billion years ago, when intense volcanism on the Moon caused hot basalt lava to pour out onto the Moon's surface. Basalt's generally dark color gives the Lunar Mare their grayish-black appearance.

The Magma Ocean and mare volcanism discoveries were certainly a victory for Dr. Noom and her colleagues, but now they only give Dr. Noom big headaches. Since she is now interested in the birth of the Moon, Dr. Noom has no choice but to try and "read" through the Magma Ocean and mare volcanism events to get to the secrets of lunar formation. Her problem is similar to the one Columbo would face if someone had tampered with the evidence before the police arrived. Columbo would have to resort to finding out as much as possible about how this individual had disturbed the crime scene. Hard? You bet, but not impossible.

Now on her second pot of coffee, Dr. Noom was visited by Dr. Ranul, a scientist just down the hall. Like Noom, Ranul had spent the last twenty years tackling the lunar origin problem, but from a different angle. While Dr. Noom studies the chemistry of rocks from the Earth and Moon, Dr. Ranul uses the laws of physics to simulate the origin of the Moon. His mathematical and computer models incorporate the physical properties of the Moon, such as its size, mass, and rotation. They also include as constraints (conditions the models must satisfy) the many conclusions about the Moon's birth that geochemists have already arrived at. The more constraints Dr. Ranul uses, the harder it is to get his models to fit. On the other hand, the more constraints the models do fit, the closer they come to potentially describing the actual lunar birth process.

Dr. Ranul was visiting Dr. Noom with the hope that she had some new data he could add to his latest model. As he poured himself a coffee from Noom's fresh pot, he mentally reviewed the models for the Moon's formation that seemed to have some chance of working. These large-scale models had been devised over a period of more than a century, but, like any good problem, it was fleshing out the details that was always the hardest. There are four main models:

1) The Intact Capture Model: The Earth and the Moon formed separately in the solar system. The Earth was going about its business orbiting the Sun when the Moon decided to pay the Earth a visit. (Actually, it was on a different path through the solar system that at some point crossed with the Earth's.) Getting too close, it got caught by the Earth's gravity and settled into its orbit around the Earth.

2) The Coaccretion Model: According to this scenario, the Earth and the Moon formed together at the same time from the same material. Before our solar system was born it existed as what is currently called the solar nebula—a rotating disk of gas and dust with a very hot core, or protosun, at the center. Some of the gas crystallized, just like the minerals in the Magma Ocean, to form mineral dust. Surface tension (the same force that holds a drop of water to a window even though it seems as though it should slide off) made particles of mineral dust stick to each other and these bigger particles were eventually attracted to each other by the force of gravity to form rock. This is the process of accretion. As more

and bigger rocks accreted together, a small Earth was formed which had even greater mass, and gravity, to attract even more matter. The theory of coaccretion states that the Moon accreted at the same time from gas, rock, and dust orbiting the accreting Earth.

3) The Earth Fission Model: In this physically dramatic model, the early Earth was molten, so it was covered with its own Magma Ocean. It also rotated very fast. At a certain point' the early Earth reached a state where its rotation was no longer stable. It has been predicted that this would occur if the Earth made one full rotation every two hours—twelve times as fast as it now rotates. As a consequence of this instability, the Earth expelled a large amount of molten liquid from its outer parts. The crystallized ball of liquid that was thrown off from the Earth became the Moon.

4) The Collision Ejection Model: This scenario envisions that the young Earth is trying to get some peace and quiet when a small planetary body, called a planetesimal collides with it. It strikes the Earth at such an angle that a large hunk of the Earth is literally vaporized. This vapor crystallizes to form dust and rock, part of which falls back to the Earth and part of which forms a ring of debris orbiting the Earth. The Moon forms by the accretion of this ring of material.

Dr. Ranul was aware that the weight of scientific opinion had shifted over the years to favor the Collision Ejection Model. Like many planetary physicists, he was currently inclined to favor it. So was Dr. Noom, although she also believed the Coaccretion Model had merit. Dr. Ranul sat down across from Dr. Noom's paper-strewn desk, took a sip of coffee and leaned forward. "OK. What do you have for me? Anything good?"

Three figures illustrating the Collision Ejection model. The images are based upon a computational model.

1. After the planetesimal collides with the Earth, a disk of material condenses around the Earth from the material vaporized during the collision. As shown in this figure, a spiral-arm structure forms as a result of the gravitational instability of the disk.

2. A lunar seed forms by accretion of material in the disk.

3. Accretion of material continues as the Moon grows in size. Lunar formation is nearly complete.

Dr. Noom had been working with some Apollo 16 samples from the Lunar Highlands. She knew that her news would be likely to change his cheery mood. "Nope, you're not going to like this," she admitted.

Dr. Ranul's jaw dropped when he saw the data. Contrary to what he had expected, these new, high-precision numbers showed no difference between the composition of the Earth's oxygen and the Moon's oxygen.

These numbers were important to Ranul because the composition of oxygen in Moon rocks serves as a powerful constraint on models of the origin of the Moon. The data compiled by Dr. Noom is called oxygen isotope data, and it is arrived at by studying the atomic composition of oxygen in rocks using a machine called a mass spectrometer. An element is defined by the number of protons it contains, and its atomic weight is determined by the sum of protons and neutrons in its nucleus. Isotopes are atoms with the same number of protons and different numbers of neutrons. All elements come in a variety of isotopes, which are named by their atomic weight. For example, two isotopes of carbon with six and seven neutrons are referred to respectively as carbon 12 (^{12}C) and carbon 13 (^{13}C), since both have the same number of protons (6). The chemical properties of isotopes of the same element are essentially the same, which means that, for example, the chemical properties of candle wax made with pure ^{12}C are essentially the same as those of candle wax made with pure ^{13}C. When rocks from the Moon were returned and analyzed, it was found that the isotopes of oxygen in lunar rocks were indistinguishable from the isotopes of oxygen found in rocks from Earth.

Why is this similarity between terrestrial and lunar oxygen important? Because evidence from meteorites shows that the oxygen composition of extraterrestrial materials depends upon where in the solar system they formed. This conclusion has recently been confirmed by analyzing Martian rocks hurled into space by asteroid impacts that eventually fell to Earth as meteorites. These Martian meteorites all have an oxygen isotopic composition distinct from that of the Earth—and the Moon. This finding was a real coup for the geochemists, but caused a real headache for the modelers. It meant that the materials from which the Earth and the Moon formed had to be nearly identical. The implication was that the Moon either had to have formed from the same rock and dust reservoir as Earth, or more tantalizingly, that the Moon was formed from matter that had actually originated on Earth.

These oxygen isotope data severely constrain the Collision Ejection Model, the one the scientists tended to favor. This theory proposes that a Mars-sized planetary body struck the

Earth and that the Moon was formed by accretion from the materials that were ejected from the Earth and/or projectile during the collision. For the Collision Ejection Model to work, and to satisfy the constraint that the Earth and the Moon have the same oxygen isotopic composition, either one of two scenarios must have occurred. The first scenario is that the Earth was hit by a projectile with a similar oxygen composition to itself. From an oxygen isotope standpoint, this would allow material vaporized from both the Earth and the projectile to contribute to the Moon (since both would have identical oxygen). A major difficulty with this scenario is that the odds of a projectile with the Earth's exact oxygen isotopic composition hitting the Earth are rather slim. The second scenario is that the Earth was struck at such an angle that the projectile did not completely vaporize and therefore did not contribute much material to the newly-forming Moon. Although this condition allows for a projectile of different oxygen isotopic composition from the Earth, it requires special impact angles that are relatively uncommon. Such angles of impact do appear to exist—we would call them glancing blows—but the problem has become much harder to solve.

Not long ago, Drs. Noom and Ranul put their heads together and realized that Dr. Noom's new technique for analyzing oxygen isotopes would allow them to test the theory that very small differences in the oxygen compositions of the Earth and the Moon actually do exist. If so, a case could be made that at least some material from a nearly-vaporized projectile with a different oxygen isotope signature from the Earth had been incorporated into the Moon. Dr. Noom's new data showed no such difference; this was the result that changed Dr. Ranul's mood this morning.

"Well, it's back to your code for you, buster," said Dr. Noom to her grumbling colleague.

"Check your numbers again," urged Dr. Ranul. "There's got to be some variation between the Earth's and the Moon's oxygen isotopes!"

Dr. Noom let that one go by without a response. She shook her head at her mass spectrometer, and got back to her numbers. Twenty miles away, Columbo was shaking his head at some pictures from the 238 Uranium Lane homicide. It was going to be a long day. ❻

Origin and Evolution of the Continents

Roberta L. Rudnick

The continental crust doesn't end at the water's edge, but extends some distance offshore, to the break in topography known as the continental slope. This steep drop-off marks the edge of the continent. Because the crust is too thick to drill through, its lower boundary with the mantle must be determined remotely, using seismic waves generated by earthquakes. On average seismic waves travel through the shallowest 40 kilometers of the Earth at approximately 6.5 km/sec.

Roberta L. Rudnick is a Professor in the Department of Geology at the University of Maryland.

Mt. St. Helens before the 1980 eruption. Nearly the top 1,000 meters of the volcano was lost in the explosion that devastated much of the surrounding area. Such explosive volcanism is typical of the water-charged andesitic and dacitic magmas that erupt in continental arcs.

Then a jump in wave speed occurs (to about 8 km/sec), reflecting the presence of material that is denser. This seismic boundary between mantle and crust, known as the Mohorovicic discontinuity, or "Moho," marks the base of the continental crust.

There are several key differences between the composition of the Earth's oceanic crust and its continental crust. The oceanic crust is predominantly thin, basaltic (that is, rich in both magnesium and iron, and depleted in silica), and young (less than 200 million years old). In contrast, the continental crust is thick, has an andesitic composition (markedly more enriched in silica, and depleted in magnesium and iron compared to a basalt), and old. The continental crust contains virtually every rock type known to geologists, including the oldest known rocks on Earth, the 4.03 billion-year-old Acasta gneisses of Canada. Thus the continental crust provides us with important clues about the Earth's earliest history and the evolution of our planet and its life forms. It's no wonder, then, that the origin and evolution of the continents has been an enduring research topic for earth scientists since the creation of the discipline.

In order to understand how continents form, we can look at where this is happening today, following the uniformitarian idea that the laws of nature do not change with time. Continental crust formation is largely accomplished by the movement of magmas, or bodies of molten rock, from the mantle deeper in the Earth towards the surface. This is therefore inherently an igneous process—one that involves the solidification of rock from a molten state. Igneous activity on Earth is related to convection in the mantle below. (To learn more about mantle convection, read the essay by

Peter Bunge in Section Two.) This igneous activity is presently concentrated in three areas:

1) Mid-ocean ridge spreading centers. Here the plates that make up the surface of the Earth are pulled apart by tectonic forces. The mantle wells upward and melts, generating basaltic magmas that solidify in the gap formed by plate separation to form new oceanic crust.

2) Subduction zones. These are sites where oceanic crust is being thrust back into the mantle. These can be located totally within ocean basins, like the Marianas island arc in the western Pacific, or at continent-ocean boundaries such as the western margin of South America, where the Andes mountains mark the combined processes of magmatism (the movement of magma) and crustal thickening.

3) Isolated points within tectonic plates, called "hotspots," such as Hawaii or Yellowstone. Here narrow upwellings of solid rock, called plumes, rise within the mantle because they are hotter and less dense than the rock around them, and create magmas that erupt on the surface of the overriding plate.

Activity at two of these three settings— subduction zones and certain hotspots— enlarge the continental crust.

Which one generates the most continental crust? The composition of the crust provides clues. The most significant clue is the fact that the composition of the continental crust—the average chemical composition from surface to Moho—is well established to be andesitic. So is the most common rock type in continental subduction zones. Moreover, the relative concentration of certain trace elements (elements present in concentrations less than 0.1 percent by weight) suggests that most of this magma was generated in subduction zone

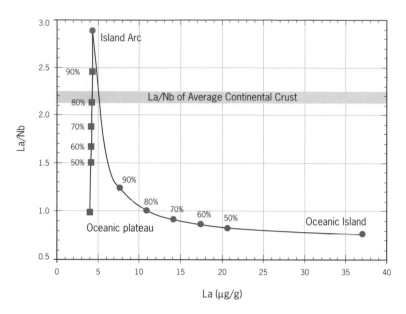

Figure 1: This plot shows the ratio of the two trace elements Lanthanum and Niobium (La/Nb) for magmas that erupt in different regions of the Earth. Magmas that erupt in subduction zones have very high La/Nb (labelled 'Island Arc'), whereas those that erupt at hot spots (labelled 'Oceanic Island' or 'Oceanic Plateau') have very low La/Nb. The average La/Nb of the continental crust, shown by the horizontal bar, is much closer to that of subduction zone magmas. This suggests that most of the continental crust was generated by subduction zone magmatism. The percentages indicate mixtures between the different magma types.

settings. For example, magmas that are depleted in the element niobium relative to the element lanthanum, which leads to high La/Nb only in subduction zones (Figure 1). This depletion of niobium relative to lanthanum is also a characteristic of continental crust.

In the early 1960s, scientists proposed the "andesite model" for the growth of continental crust, based on two observations. The first was that continental subduction zones are marked by the spectacular eruption of andesitic magmas (such as Mt. St. Helens, Washington and Mt. Pinatubo, Philippines). The second was that the continental crust is generally younger at its edge and grows progressively older towards the interior of the landmass. According to the

"andesite model," continental growth occurs because of the addition of andesitic magma, either at subduction zones at the margins of continents or when andesitic island arcs collide with continents. Island arcs are generated when one oceanic plate is subducted beneath another, and when the oceanic plate carrying the island arc is subsequently subducted beneath a continent, the island arc collides with a continent, thereby increasing continental mass.

However, there are two fundamental problems with this elegant model:

The first problem relates to the types of rock found in very old continental crust. A significant volume of the crust probably formed before 2.5 billion years ago, in the Archean Era. Yet andesites are scarce in Archean rock formations, which tend to be dominated by basalts and granites. The scarcity of andesites in these old rock formations suggests that the "andesite model" did not apply to crust formation in the Archean.

The second problem is that andesites cannot be generated directly by melting the mantle. Typically, when the upper mantle melts to produce a magma, and the magma moves upwards and solidifies, the rock that forms is basaltic. In order to become andesitic, the magma must go through a process called crystal fractionation. Crystal fractionation is a process that occurs when magma cools. Crystals form that are different in chemical composition from the original molten rock. If these crystals are removed from the magma, the composition of what remains—the derivative, or residual magma—changes, too. The crystals that form from a basaltic magma are rich in iron and magnesium; the magma that's left after their removal is andesitic. The crust can also become andesitic when the basaltic magmas from the mantle mix with magmas formed when certain kinds of rocks in the pre-existing crust, such as granites, melt.

The fundamental paradox remains: basalt is the primary product of mantle melting, yet the continental crust is not basaltic. If andesites are scarce in older rock formations and cannot be generated by melting the mantle, how did the crust come to be andesitic? Additional processes must have played important roles in producing the continental crust as we see it today.

To explain this paradox, scientists have suggested four distinct processes:

1) Heavy, or high-density, iron- and magnesium-rich crystals are separated from basaltic magmas by crystal fractionation near the base of the continental crust. The crystals sink from the base of the continental crust into the mantle, in a process called delamination. The crust that remains is andesitic.

2) Silicic magma (more enriched in silica and depleted in magnesium and iron than an andesite) is produced within the mantle by melting slabs of basaltic oceanic crust as they are thrust deep into the Earth at subduction zones. Scientists think that slab melting is unlikely to occur in most modern subduction zones, but believe it could have been an important process earlier in Earth history. At that time, mantle temperatures were significantly higher because radioactive, heat-producing elements—potassium, thorium, and uranium, much of which have since decayed away—were more abundant. Silicic igneous rocks that bear chemical fingerprints of having been formed by slab melting are abundant in most Archean rock formations. When combined with the basalts that are also present in these formations, these silicic magmas generate an average andesitic composition of the Archean continental crust.

3) Weathering changes the chemistry of the rocks in the continental crust. Weathering strips out the soluble elements and transports them into the oceans. Most of these elements, such as calcium and sodium, accumulate in sedimentary rocks in the ocean that are then moved back to the continents by tectonic activity. Magnesium, on the other hand, is recycled into the mantle by hydrothermal activity at mid-oceanic ridges and subduction of the altered oceanic crust. The net result of this chemical cycling might be to create crust with an andesitic composition from crust that was originally basaltic in composition.

4) As discussed above, crystal fractionation causes the basaltic magma to differentiate into two components: iron- and magnesium-rich crystals, and andesitic magma. Because these crystals are high-density, they could accumulate beneath the Moho and therefore be interpreted as mantle. What remains above the Moho, which scientists choose to define as the base of the continent, are rocks with an andesitic composition.

These processes are not mutually exclusive, and all probably contributed to shaping the continents as we see them today. However, some may have played a more important role than others. Most continental igneous rocks do not show the chemical signature of weathering. Nor have large masses of dense accumulated crystals been found beneath the Moho in most continental regions. This suggests that the first two—delamination and slab melting, with slab melting occurring more significantly early in Earth history—are the most significant processes in producing the buoyant, andesitic crust upon which we live.

Continents are one of the features that make Earth unique in our solar system. Why are continents apparently restricted to the Earth? Many planets have ancient crusts still pockmarked by the intense meteorite bombardment that occurred early in the solar system's history. All vestiges of Earth's original crust were completely destroyed by the subsequent movements of tectonic plates, which were probably more vigorous early in the planet's history. The action of plate tectonics doesn't occur on other planets, and the igneous products of plate tectonics—subduction zone magmas—are the basic building blocks of the continents we live on.

Mt. St. Helens, two years after the eruption.

Life and the Evolution of Earth's Atmosphere

Stephen J. Mojzsis

What's Life to a Geochemist?

Thinking about it, there's a subtle kind of poetry involved with its scientific definition. Life can be said to be a self-replicating, encapsulated, chemical system that undergoes Darwinian evolution. In other words, groups of related organisms have evolved in response to changes in the environment through natural selection. This process has been in effect ever since life began from simple organic molecules in water, over four

Stephen J. Mojzsis is an Assistant Professor in the Department of Geological Sciences at the University of Colorado, Boulder.

billion years ago. So, the origin of life is probably the origin of evolution. Life is resourceful and entrepreneurial. It takes advantage and it changes the chemistry of its surroundings. Life is a fantastically complex system, the emergence of which remains the greatest mystery in science.

Long-term changes in the composition of the atmosphere and oceans are intimately linked to both the geophysical changes in the solid Earth itself and with the ongoing evolution of life. The atmosphere and oceans first appeared about 4.5 billion years ago, soon after the Earth and Moon completed their formational phase. This was the time when gases escaping through volcanoes made an envelope of atmosphere around the young Earth, and a primitive crust solidified and cooled to the point where liquid water could condense. Water began pooling into the first lakes, seas, and oceans. The interaction of water, heat, and rock set the stage for the origin of life.

From the beginning, a number of factors have affected the makeup of the atmosphere to change it from its initial state to what we have today. Several of these factors, such as plate tectonics, weathering (which recycle rocks, water, and gases), and chemical changes induced by the byproducts of life itself, are internal to the planet. However, external factors such as the slowly but ever-increasing luminosity of the Sun over billions of years, gradual changes in the Earth's orbit over many tens of thousands of years, and the rare but catastrophic impacts of giant meteorites and comets, have also played an important role.

The atmosphere of our planet did not originally contain all the free, breathable oxygen it does now. The first permanent atmosphere arose when gases that had been dissolved in the molten planet during its assembly from smaller bodies, called "planetesimals," were released to the surface by volcanism. That first, primitive atmosphere was probably several times denser than what we have now, and was dominated not by oxygen, but by carbon dioxide—a major greenhouse gas. Other gases, such as molecular nitrogen, water vapor, and small amounts of carbon monoxide, sulfur gases, and trace quantities of methane, and hydrogen were also present.

Astrophysical computer models based on the study of young stars and of star-forming regions in the galaxy strongly suggest that the Sun was much dimmer when the first life emerged on Earth, over 4 billion years ago. A dimmer Sun would have supplied less solar radiation to warm the early Earth. To keep the Earth from starting out as a frozen wasteland, with no hope for beginning life to take hold, an atmospheric "greenhouse" must have kept the surface zone warm enough to maintain water in liquid form. Liquid water is the prerequisite for life. Greenhouse gases in much higher abundance than today, primarily water vapor, carbon dioxide, and methane, would have formed a thermal blanket over the surface of the early Earth that strongly absorbed outgoing thermal radiation. This leads to a significantly enhanced, warming "greenhouse effect" that offset the dimmer Sun. Without this very different, greenhouse gas-rich and oxygen-poor early atmosphere to begin with, life would have gotten a frozen start. (See the essay by Charles F. Keller in Section Five, which discusses greenhouse gases and global warming.)

The composition of this early atmosphere, so different from the one we have now, would be deadly for most life that is not a primitive bacterium. Soon after the emergence of the first life more than 4 billion years ago, the activities of organisms began to influence the composition of the atmosphere. As Earth's biosphere and atmosphere co-evolved over billions of years, the product of photosynthesis, mainly, free oxygen, has come to dominate the chemistry of the atmosphere. Without the byproducts of crucial biological processes such as photosynthesis, there would be little or no free oxygen for animals to breathe, nor enough to form a protective ozone layer in the upper atmosphere. Ozone (O_3) is formed by the recombination of oxygen (O_2) by solar radiation in the upper atmosphere into the new molecule with three oxygen atoms. This molecule screens harmful ultraviolet sunlight and prevents most of it from reaching the surface. Without an effective ozone layer in the upper atmosphere, more advanced life would not have been able to emerge from the seas to colonize the land. We can safely say that the vast amount of ozone needed to permit the colonization of land became possible only after significant free oxygen accumulated from photosynthesis beginning early in Earth's history.

Since little evidence from the Earth's formational time has survived for scientists to study, it is possible to only make a few basic conclusions about the nature of the first atmosphere and the conditions in which life originated. Most of what is known about early planetary history comes from three sources: the sparse geologic record remaining from more than 3.5 billion years ago, computer models of atmospheres changing with time, and comparing Earth to its planetary neighbors, Mars and Venus, which evolved along very different paths. Scientists have also recently started to study the sophisticated biochemical machinery of primitive photosynthetic bacteria to learn more about their early evolution. These primitive bacteria are descended from the pioneering microorganisms that forever changed the chemistry of the atmosphere from a primitive unbreathable mix to oxygen-rich air.

How Earth's Atmosphere Originated

The elements and compounds that make up the atmosphere and oceans evaporate readily under normal surface conditions; that's the definition of "volatile." Elsewhere in the universe, these volatiles would assume very different physical states. In the space between planets and stars, where it is very cold, they would be present mostly as solid ices; however, near stars or in regions where new stars are being formed, where it is hot, they would be present as plasma.

Plasma is a fourth state of matter after solid, liquid, and gas. It exists only at extremely high temperatures when electrons are stripped off atoms to form charged particles. Plasma is in fact the most common state of matter in the universe, yet this state is useless for life; only solid bodies like planets can support matter in the usable solid-liquid-gas states. Life as we know it can occur only where a volatile such as water is stable in the liquid state, and where energy resources like sunlight or chemical energy are available for exploitation for making or supplying food. Such special conditions for life seem to be satisfied only on planetary

Table 1: Comparison of atmospheric compositions of Venus, Mars, Earth						
	pressure (bars)*	CO_2 (%)	N_2(%)	Argon-36 (%)	H_2O (%)	O_2(%)
Venus	92	96.5	3.5	0.00007	<0.00003	trace
Earth	1.013	0.033	78	0.01	<3	20
Mars	0.006	95.3	2.7	0.016	<0.0001	trace

*A bar is a measure of pressure. One bar equals 1.013 atmospheres, or the atmospheric pressure at sea level.

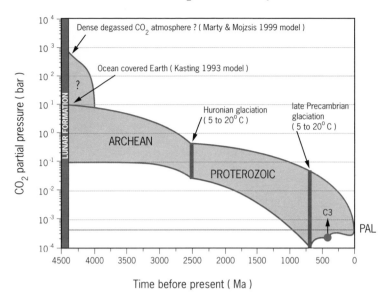

Partial Pressure of CO₂ in the Atmosphere Over Time

Figure 1: Changing concentrations of atmospheric carbon dioxide with time on Earth in response to the steady increase in solar luminosity (after Kasting 1993). The amount of past CO_2 that has been calculated in this figure (using the model of J.F. Kasting, 1993) is the concentration required to keep the surface warm enough for liquid water to exist even with past lower solar output. PAL = present atmospheric level of CO_2. The Moon formed between 4.5 and 4.45 billion years ago (dark blue field). The Earth could have started with an atmosphere extremely enriched in carbon dioxide (gray field). The rise in land plants after 500 million years ago ("C3" on the figure) defines a minimum limit for CO_2 in the air after that time.

surfaces. On Earth, volatiles that constitute the atmosphere are principally represented by the gases nitrogen, which makes up seventy-eight percent, oxygen (twenty percent), water vapor (percentage varies according to weather and geography), argon gas (about one percent), and carbon dioxide, or CO_2 (about 0.03 percent), and trace amounts of other gases. This mixture is vastly different from the lifeless atmospheres of nearby Venus and Mars, both of which have no free oxygen and are completely dominated by carbon dioxide (Table 1). This comparison of the atmospheres of our planetary neighborhood illustrates the role of life in maintaining surface conditions amenable for habitation.

The air and oceans probably represent most of the surface reservoir of the volatile compounds Nitrogen (N_2) and water (H_2O) on Earth. Oxygen is dominantly locked in minerals in the crust and the Earth's interior, and would remain so entirely if not for the actions of photosynthesizers. As for carbon, the amount that is present as CO_2 in the atmosphere is small compared to the majority of the Earth's carbon which is sequestered in carbonate rocks such as

limestone and marble, organic matter in sediments, fossil fuels, and biomass on or near the surface. If all of this carbon were oxidized into CO_2 and put into the atmosphere, it would overwhelm all other gaseous components. The amount of CO_2 released in such a case would give our planet a surface pressure of about sixty bars, and result in an atmospheric composition very similar to that of Venus. There would also be a massive greenhouse effect, like the one that keeps the surface of Venus so hot (greater than 460°C). This massive greenhouse effect would result in temperature increases that were more than enough to vaporize the oceans. If the enormous volume of water in the ocean were vaporized, a heavy (250 bar) and deadly greenhouse blanket would push temperatures up beyond even the melting point of rocks. Therefore it is easy to understand how a delicate balance is necessary among the biosphere, atmosphere, crust, and the oceans to keep the Earth habitable over long time scales.

Over the past few hundred million years, the amount of CO_2 in the atmosphere has decreased from high levels to stabilize near its present relatively low level (Figure 1). This appears to be a response of the environment to the steady increase in solar luminosity, which has kept the surface within an equitable temperature range for life to flourish over time.

Photosynthesis Produces the Oxygen in the Atmosphere

Photosynthesis is the process through which certain bacteria and plants use carbon dioxide, water, and light energy to make food and oxygen. The first life on Earth was probably not photosynthetic, as conditions on the surface were harsh because of multiple meteorite impacts early on, and a deadly bath of ultra-violet radiation before the establishment of an ozone screen. Without an ozone layer, there might still be a biosphere, but it would be tiny and struggling to eke out an existence deep in the crust away from the surface. The bottom of the sea was protected from the realities of a dangerous surface zone on the young Earth, so life might have originated and subsequently evolved there, only to later migrate upwards. In the darkness of the ocean depths, it would have had to rely on chemical rather than light energy. An environment where this is possible—the deep-sea hydrothermal vent environments—has existed throughout much of Earth's history along the mid-ocean ridges (to learn more about this

environment, read Deborah S. Kelley's essay in Section Six). We don't know how long it took for photosynthetic life to evolve and begin the production of oxygen, but we do know that photosynthesis is the dominant metabolic process that sustains the biosphere today.

Photosynthesis, although a cornerstone of the present biosphere, is a complex process that as yet remains poorly understood. In order to convert light energy into chemical energy, photosynthetic organisms use molecular pigments that absorb sunlight of different wavelengths and reflect others. Different colored molecules in plants absorb light energy from different parts of the solar spectrum. The principal light-collecting pigment in plants is chlorophyll, which gives them their green color. Interestingly, the Sun's output is dominated by blue-yellow light, making this the most intense part of the visible solar spectrum. So, chlorophyll is green because it absorbs all of the intense, visible, blue-yellow light and reflects the green. Plants, like all forms of life, are efficient, so they make special use of their pigments to reflect the least intense parts, including the green light of the solar spectrum. The intense light energy is used by the chlorophyll in plants to make carbohydrates or food, with oxygen as a waste product.

How the Biosphere Affects the Atmosphere

Organisms affect the composition of the atmosphere in several significant ways. Most of us are aware that animals inhale oxygen and exhale carbon dioxide when they breathe. These gases are exchanged between the biosphere, the atmosphere, the hydrosphere (oceans) and the lithosphere (crust) in enormous quantities. Only a tiny proportion of the oxygen in the atmosphere is made when ultraviolet light splits water molecules high in the atmosphere in a process known as photolysis. The vast bulk of oxygen is continuously replenished via plant photosynthesis. The high concentration of oxygen needed to keep the contemporary biosphere going, including all of the humans breathing it, is maintained despite its continual removal through oxidative processes like weathering of rocks (which tend to give them a reddish color as in the Grand Canyon), rusting, burning, and so on. In fact, if all photosynthesis were to cease, which would be a catastrophic event indeed, the oxygen in the atmosphere of the Earth would be mostly gone in less than a million years, along with the protective ozone layer. In the absence of photosynthesis, the oxygen present in the atmosphere would combine with carbon in organic matter at the surface to make carbon dioxide, or react with iron in minerals, to soon be removed from the atmosphere altogether. In 1979, in his book *Gaia: A New Look at Life on Earth,* James Lovelock emphasized that the atmospheric composition of the Earth is far from achieving chemical equilibrium with the surface and must be sustained by biological processes. According to Lovelock's Gaia Hypothesis, life is responsible for making the unique conditions for habitability on Earth; life regulates the global environment. The other planets near the Earth provide a useful measure of how different our planet is from places that are in equilibrium. Again, take a look at the lifeless planets of Mars and Venus in Table 1. They have negligible amounts of oxygen in their atmospheres, and are dominated by carbon dioxide. The same would be true for the Earth if life had never appeared and changed the atmosphere to keep the planet habitable.

The Build-up of Oxygen
When oxygen began to first accumulate on Earth billions of years ago, it caused a crisis for primitive anaerobic bacterial life, which is actually poisoned by excess oxygen in the environment. Such life still exists today in organic-rich soils, swamps, deep in sediments, and lake bottom muds where levels of oxygen are low. This early oxygen crisis, however, created an opportunity and led to important new evolutionary changes as life responded to this challenge with metabolic innovations that made use of the newly-available oxygen. Organisms evolved which used oxygen in their metabolic cycles to get more energy out of their food—twenty times more than anaerobic organisms get.

A coupled biological and geological mechanism termed the "carbon cycle" began to manifest itself early in the history of life on Earth, as life took over the production of organic matter and oxygen and began to regulate the balance of carbon between the atmosphere and the oceans. This cycle has helped to regulate the planet's surface temperature by balancing CO_2 output from volcanoes with weathering, burial of organic matter in sediments, and other means of removing CO_2 from the atmosphere. (To learn more details about how this works, read Rachel Oxburgh's essay on the carbon cycle in Section Six). In a sense, industrialization and the burning of fossil fuels, which have released vast amounts of carbon dioxide into the atmosphere, represent a small reversal of the general trend toward higher oxygen concentrations and lower carbon dioxide concentrations with geologic time. Human activity, the long-term actions of plate tectonics, and variations in Earth's orbit will inevitably cause large fluctuations in the amount of carbon dioxide in the atmosphere again, and perhaps changes in oxygen concentration as well.

When Did These Changes Occur?
Life originated more than 4 billion years ago. How soon after the origin of life did oxygen levels begin to rise? This is a question that scientists have been exploring for some time (Figure 2). Data gleaned from ancient soils

(paleosols) and a variety of other geological evidence indicates that starting about 2.5 billion years ago, sedimentary rocks began to become increasingly affected by higher amounts of oxygen in the atmosphere, and appear red, as though "rusted." Scientists believe that oxygen concentration began to rapidly increase at that time because more oxygen began to be produced by photosynthesis than could be whisked away by reactions with the atmosphere, oceans, and rocks. A threshold had been reached, and the rise of oxygen had begun. At around that same time, more complex bacterial life evolved in response to the oxygen-rich conditions, including organisms with a true cell nucleus (Eukaryotes) from which all plants, animals, and fungi are descended.

The final significant section in the emergence of an oxygen-rich atmosphere on Earth began about 700 million years ago, when the fossil record documents the emergence of multicellular organisms. When oxygen levels became high enough, a little more than one percent of the present level, a powerful ozone screen started to form in the upper atmosphere that protected the first land-dwelling creatures from the Sun's harmful ultraviolet radiation. Abundant oxygen in the atmosphere was essential to the rise of animal and plant life on land.

A Hot Future
Knowing what we do about the past trends in atmospheric evolution of the Earth, what can we say about the distant future for life on this planet? The Sun will continue to grow brighter over the 5 billion years or so that remain in its stellar life span. Its luminosity has been increasing on a near-linear path for more than 4.5 billion years because hydrogen is continually converted into the heavier element, helium, by nuclear fusion. This conversion releases tremendous amounts of energy every second and is responsible for sunlight. As

helium accumulates, the average density of the Sun increases over time, which in turn causes pressures and temperatures in the Sun's interior to increase. These higher pressures and temperatures increase the rate of fusion, and the extra energy that is generated leads to increased luminosity. As this positive feedback continues, the Sun will continue to increase in intensity by about six percent or more in the next billion years.

One of two things needs to happen to keep the Earth from slowly going the way of deadly hot Venus. Either almost all of the greenhouse gases have to be removed from the atmosphere, or the planet's albedo—the amount of sunlight the surface can reflect back into space—must increase. Unless new plants evolve a type of photosynthesis that requires very little carbon dioxide, or some future culture somehow protects the Earth from a brighter Sun, life on land will have to be abandoned. In the end, some 5 billion years from now, much of the Sun's fuel will be exhausted. The Sun will begin to grow as it approaches the Red Giant phase of its stellar life cycle. By then, the oceans on Earth will have long since boiled away. The last life, primitive bacteria much like the first that appeared, residing in deep crustal rocks and sediments, will have perished. As the surface temperature rises, all of the volatile gases stored over many eons in the Earth's crust will be released to form a thick, hot, blanketing atmosphere. Finally, there will be twin Venuses from that time until the end of the solar system.

Summary
A great deal can be learned by approaching the history of life on Earth from the standpoint of geology. The rocks that make up the geologic record preserve changes in the planet's crust, oceans, and atmosphere that have accompanied the history of the evolution of life. Studies of the geologic record

help us fine-tune theoretical models that attempt to shed light on conditions at the Earth's surface when life emerged. Studying neighboring worlds is also relevant because it allows us to compare the very different paths taken by these planets in the absence of life. It is also necessary to keep in mind the immense expanse of time involved. Geologists approach the geologic record much like a detective tackles problems in a detective story; the plot thickens as new evidence is uncovered, but it often yields more questions than answers.

In science, theories evolve as evidence accumulates. But theories must be grounded in the solid principles of physics, mathematics, and chemistry that describe the behavior of matter in the universe. We are now on the threshold of being able to evaluate the place of Earth in the universe; we may discover that we are not all that unique in having a water-rich planet with abundant life and oxygen. The search for other worlds that contain, or contained, life stands as the grand quest for the next millennium.   ☉

Partial Pressure of O$_2$ in the Atmosphere Over Time

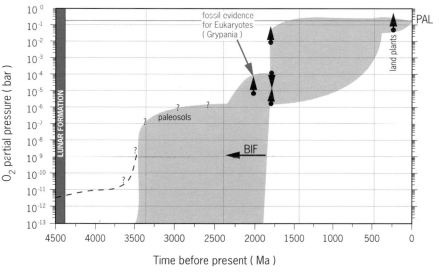

Figure 2: This plot (after Kasting, 1993) shows estimates of changing concentrations of free oxygen in the atmosphere over geologic time. The area (in light blue) represents the range in possible concentrations of oxygen based on model calculations, and the study of ancient soils (paleosols), fossil organisms, and marine sediments that only form in the absence of oxygen. These sediments are preserved as Banded Iron Formations (BIF) and only appear in the geologic record up to about 1.8 billion years ago. Although it is not known when photosynthesis began, it is clear that photosynthesis only became an important producer of oxygen in the atmosphere late in Earth's history. PAL = present atmospheric level of O$_2$.

Camels in the Mauritanian desert.

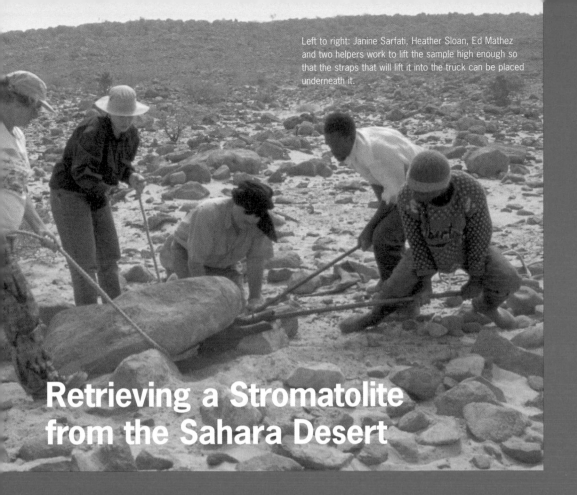

Left to right: Janine Sarfati, Heather Sloan, Ed Mathez and two helpers work to lift the sample high enough so that the straps that will lift it into the truck can be placed underneath it.

Retrieving a Stromatolite from the Sahara Desert

How do earth scientists prepare for a two-week expedition to Saharan Africa? "First of all, we got vaccinated," replies Heather Sloan ruefully. A Research Scientist and Exhibition Coordinator for the Gottesman Hall of Planet Earth, Dr. Sloan became a geologist because, "I was always curious about the world around me. I loved getting information that I could use to answer questions like, why is that rock here? Or why is the land shaped the way it is?" Curiosity, a Ph.D. in marine geophysics, many shots, and twenty hours in the air landed Sloan in Nouakchott, Mauritania in February 1998. There, she and the Museum team, which included Edmond A. Mathez, head of the Department of Earth and Planetary Sciences, were met by Dr. Janine Sarfati of France's Université de Montpellier 2, who knew the region well. The purpose of the two-week expedition was to bring back a rock called

a stromatolite for display in the Museum's Gottesman Hall of Planet Earth. This is a new exhibit built around dramatic rock samples that exemplify the dynamic processes that shaped the Earth.

Stromatolites are "organo-sedimentary" structures—basically slimy masses—built by microbes, the only life to exist on Earth until about a billion years ago. "They are the earliest evidence of life from the Proterozoic Era, which was supposed to be 'azoic'"— without life—points out Dr. Sarfati. In addition, stromatolites played a tremendously significant part in the evolution of the ancient Earth and its atmosphere. Originally, the atmosphere contained very little oxygen. "But about 2.6 billion years ago there was a global expansion in the number of stromatolite colonies," explains Sloan, possibly because that Era's shallow coastal and inland seas provided an optimum environment for these colonies of algae. "The fact that these organisms photosynthesize totally changed the chemistry of Earth's atmosphere, because it contributed to a very rapid increase in the concentration of oxygen," she continues. "It shaped the type of life which followed, because life could become aerobic"—that is, able to consume oxygen for survival. (Steve Mojszis's essay in this section describes this process in detail.)

Stromatolites also provide information about the climate conditions under which they formed. Because most stromatolites need sunlight, their growth is generally restricted to the ocean's photic zone, typically water less than 150 meters deep, "so we know that wherever they were growing the sea must have been relatively shallow." Different stromatolite shapes reflect different water conditions as well. "If they were agitated in surf zones, stromatolites show protective 'walls' around the structures and pieces of rock in the buildup," says Sarfati. Others stand high above the sediment, which indicates that they developed in quiet water beneath the wave action.

Sarfati began studying ancient stromatolites along a 1200-kilometer long outcrop in Algeria and Mauritania in the 1960s. When French soldiers on camelback discovered the structures in the 1920s, they at first mistook them for petrified forests. Unlike the many places where stromatolites have been buried by tectonic activity, this two- to three-hundred-square-kilometer area of West Africa has remained quite flat and very stable. Consequently, it's home to unique formations of almost twenty-five different types of stromatolites—"some cone-shaped, some with branches, some shaped like big loaves of bread, called bioherms," says Sloan. The formations are composed of layers built up over

many millions of years, and are between ten and thirty meters thick and up to ten meters high. In the Sahara the formations are exposed by the outcrop and free of vegetation, so, says Sarfati, "in one glance you can trace the story occurring around you."

After a grueling drive inland to the town of Atar, it was up to Sarfati to explain the story behind different formations to the American team. "Janine has this tremendous store of knowledge and field experience about stromatolites, and along with all that, she is absolutely indomitable," recounts Sloan admiringly. "At seventy she was scampering up and down rock faces in 115°F heat. And afterwards, while most of the younger members of the party collapsed, she would drag me out into the market to look for bargains." It took three days to explore the outcrops and pick out the right stromatolite for the exhibit. Scientific, aesthetic, and logistical criteria all had to be satisfied. "We wanted one which showed the structure as completely as possible, because a lot were badly weathered. And we wanted as large a rock as possible, but we didn't have any heavy machinery, and the roads are terrible."

The boulder selected, a Jacutophyton that Sloan describes as "a central conical structure with beautiful branches coming off all around it. They grew quite close together, a bit like a submarine forest." The stone had a beautiful, iridescent desert patina, caused by "the combination of windblown minerals and the action of bacteria that live on whatever the wind brings them." It also had a delicate texture of beautiful scallop-shaped ridges built up by algae and bacteria and exposed by weathering. Fortunately, the stromatolite was already free of the outcrop and lying by the road. But it weighed 760 kilos, and it was quite fragile.

"So Janine and I went to the market and we started buying things to wrap a rock up in," recounts Sloan, grinning at the memory. Since many commodities are used and reused in the desert, rope was available only in short pieces. They found cargo nets, and rice sacks to keep the rock dry, and twine, and purchased a handmade, six-inch needle from an ironmonger. "By this time we'd acquired quite a following of children, because nobody could figure out what on Earth we wanted these things for. We also had to communicate to the guy at the garage that we wanted the totally ruined tires, the ones that simply couldn't be patched any more, to use as padding. Then we headed back to the hotel and started sewing sacks together."

A local merchant lined them up with a big truck and giant frame with a big pulley on it, but halfway to their destination the ancient vehicle broke an axle in the sand. "So a little pick-up truck passing by, the local bus in effect, agreed to take this big iron frame, and drive it out to the site—along with all of the passengers. They were all willing to come out and watch because they simply couldn't believe we were going to all this trouble to pick up a rock," recalls Sloan with a laugh. "And of course everybody participates, so we suddenly had a crew of eighteen or nineteen people setting up this huge frame. And thank goodness, because it was heavy." The team had already laid out the packing material in their little Toyota truck, the stromatolite was loaded onto the truck bed, and they trussed the whole thing up. "It took about twelve hours to get back to Nouakchott," recounts Sloan, "and it was a very exciting trip." 🌍

Left to Right: Heather Sloan, Janine Sarfati, and Ed Mathez discussing the most promising sample locations while studying a local geologic map.

42 | 43

Case Study

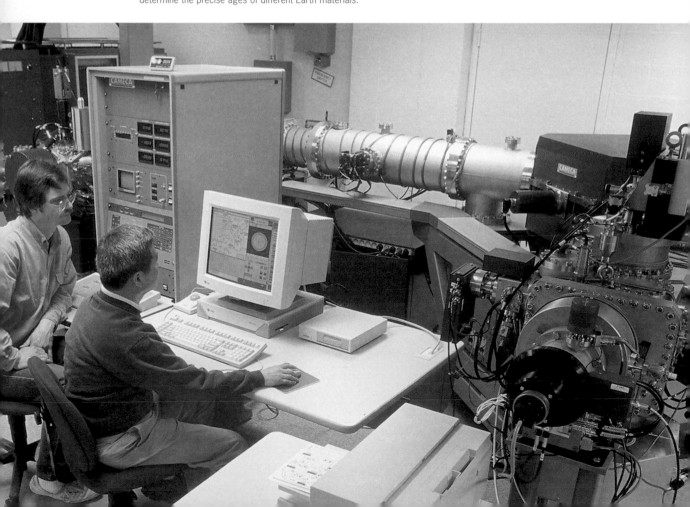

Scientists at work in an ion probe lab. The instrument shown is similar to one of those used by Meuller to determine the precise ages of different Earth materials.

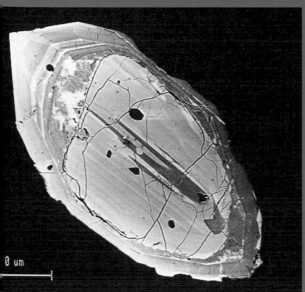

Backscatter electron image of a zircon crystal showing narrow growth zones around a central core.

0 um

Zircon Chronology: Dating the Oldest Material on Earth

What are the oldest rocks on Earth, and how did they form? The material that holds the greatest insight into these fundamental questions, because it can contain a record of some of the earliest history of the Earth, is a mineral named zircon. For example, a few grains of zircon found in the early 1990s in a sandstone from western Australia dates back 4.2–4.3 billion years, and we know from meteorites that the Earth is not much older at 4.56 billion years. Geology professors Darrell Henry of Louisiana State University and Paul Mueller of the University of Florida are expert practitioners of several techniques that can extract precise age information from zircons. They're searching for some of the oldest rocks in the continental crust, for the zircons within them, and for the clues the zircons contain about the formation of the planet.

Originally formed by crystallization from a magma or in metamorphic rocks, zircons are so durable and resistant to chemical attack that they rarely go away. They may survive many geologic events, which can be recorded in rings of additional zircon that grow around the original crystal like tree rings. Like a tiny time capsule, the zircon records these events, each one of which may last hundreds of millions of years. Meanwhile, the core of the zircon itself remains unchanged, and preserves the chemical characteristics of the rock in which it originally crystallized.

Zircon contains the radioactive element uranium, which Dr. Mueller calls "the clock within the zircon" because it converts to the element lead at a specific rate over a long span of time. According to Mueller, this makes

Darrell Henry overlooking Quad Creek in the eastern Beartooth Mountains, Montana.

zircons "the most reliable natural chronometer that we have when we want to look at the earliest part of Earth history." He goes on to explain that there are two ways to tell time in geology. "One is a relative time, meaning if there's a mineral of one kind, and growing around it is a mineral of a second kind, you know the inner mineral formed first, but you don't know how much time elapsed between the two." Henry evaluates these kinds of mineral relations in rocks. From the types of minerals and their distributions in the rocks he reconstructs a relative sequence of events that reflects the change over time of parameters like pressure, temperature, and deformation. "If I have a metamorphic rock," elaborates Dr. Henry, "I can use the types of minerals and their chemistry to determine the conditions that the rock had experienced at some point in its history. For example, a temperature of 700°C and high pressure of several thousand times atmospheric pressure imply that it had been deep in the crust at some time during its geologic history." He infers what has happened to the rocks, but

not how long ago it happened. That's where the second kind of time comes in: absolute as compared to relative. "We try to supply the when," explains Mueller. "My job is to look at the chemistry of the rock, including its isotopes, and try to derive the absolute times for events that are recorded in the rock and its zircons."

How precise are those actual numbers? "Depending on the history of the rock, we can date things nowadays down to something on the order of a few hundredths of a percent of its age," answers Mueller. That translates, for example, to plus or minus a million years out of three billion. Carbon-14 dating can go no further back than about 70,000 years, because the half-life of carbon-14 is only 5,730 years. (The half-life is the time it takes for half of the original radioactive isotope to change to another element.) In comparison, the half-life of the radioactive uranium 238 isotope is 4.5 billion years, which makes it useful for dating extremely old materials.

Zircon chronology begins in the field. "You go out and look for relative age relationships, see which rock unit was formed first," says Henry. "For example, there may be a granite which contains pieces of other types of rocks enclosed in the granite. Because of their position, we know that the rocks enclosed in the granite have to be older." Geologists map an area to identity these relative age relationships. Then they collect samples, which weigh from two to more than one hundred pounds, depending on the rock type. Zircons aren't rare; in fact, they're common in granitic rock. But they are tiny grains that make up only a small fraction of any given sample, typically less than a tenth of one percent, and they're dispersed throughout the rock. This makes separating out the zircons a painstaking process. The rock is ground up to break it into individual mineral

Paul Mueller with student Stephanie Ingle by a cliff in Stillwater, Montana.

grains. Then, "because zircon is more dense than almost any other mineral, we put the ground-up rock in a liquid with very high density so that only the densest minerals fall through to the bottom," explains Henry. In other words, says Mueller, "zircons sink. We also use the magnetic qualities of the zircons to separate the most pristine ones from the rest."

Then the detailed geochronology work begins. "I'll take a fraction of those zircons, make thin sections of them—slices of mineral thirty micrometers thick, roughly as thick as a hair, that are mounted on glass—and get an idea of what they look like in terms of zoning pattern, whether they underwent multiple episodes of growth, how simple or complex they are," says Henry. He passes this information along to Mueller, along with the sample's geological context. "I also look at a thin section of the rock to learn something about the framework in which the zircon occurs. Is it in a granite? Or is it in a metamorphic rock that has had a more complex history? Or is it a metamorphosed

sedimentary rock? By knowing its history, we can interpret the age of the rock much better."

"To understand the relative geologic history of a rock, Darrell uses thin sections because he's interested in the relations among all the minerals, which make up the rock," explains Mueller. "However, for geochronology, we're interested in the minerals that make up one tenth of one percent or less." He looks at the zircon using various techniques—"light reflected off the grains, light transmitted through them, cathodoluminescent light resulting from hitting the zircon with an electron beam"—to establish the scale at which the zircon grains should be analyzed. Quantitative microanalysis of the elements in zircon is done with an electron microprobe. "This allows us to analyze things on a micron (a millionth of a meter) scale using a thin beam of electrons," explains Henry. "The electrons irradiate the sample, causing atoms within the sample itself to give off X-rays. Each of the atoms of the different elements in the sample gives off X-rays with characteristic wavelengths. You can then compare these to a standard with a known concentration of the element, and come up with an exact composition of that small spot. An individual zircon grain may be composed of many zones of different compositions and ages. Isotopic compositions can be determined with an ion probe. Do we want to look at the whole grain, or should we direct a tiny beam of oxygen ions, 300 micrometers in diameter, on parts of the zircon grain to analyze for U (uranium) and Pb (lead) isotopes so we can date that spot and dissect the zircon's individual history?" Alternatively, the uranium and lead can be separated chemically when an individual zircon grain is dissolved in hydrofluoric acid. "Then we analyze them on a mass spectrometer, which gives us the ratios of the individual uranium and lead isotopes,

Case Study

An outcrop of the 2.7 billion year-old Stillwater Igneous Complex in the Beartooth Mountains in Montana.

and from that we can calculate the time," explains Mueller.

Ultimately, says Henry, "all of these data are combined into a larger picture of how the Earth worked billions of years of years ago." In Mueller's words, "it boils down to the fact that the more we know about the variety of rocks that made up the earliest continents and how these continents evolved, the better our window onto how the Earth formed and the early processes that separated the crust from the mantle and probably even the mantle from the core." Mueller describes his and Henry's collaboration as a parallel journey. "Our research marches down the same road, and sometimes we hold hands and sometimes we go our separate ways." In either case, they're constantly exchanging information yielded by their different approaches, and there's always something new to look at. Mueller sums it up: "One rock's a lot of work."   ☉

Arthur Holmes: Profile

Harnessing the Mechanics of Mantle Convection to the Theory of Continental Drift

English geologist Arthur Holmes made not one but two major contributions to our understanding of how the Earth works. He was the first earth scientist to grasp the mechanical and thermal implications of mantle convection, and he widely applied the newly-developed method of radioactive dating to minerals in the first attempt to quantitatively estimate the age of the Earth.

Holmes was fortunate that the phenomenon of radioactivity was discovered during his years as a graduate student at London's Imperial College of Science. Holmes had come there to study physics, but switched to geology before graduating in 1910. Meanwhile, in 1905, English physicist Ernest Rutherford had suggested that the energy emitted by radioactive minerals in the form of particles and rays could be used to date the minerals. Called radioactive dating, this technique measures the rate of decay of certain unstable atoms, such as uranium, contained within minerals. Using this new technique, Holmes was able to determine the age of minerals and thus the rocks they are in, and in 1913, he formulated the first quantitative geological time scale. He estimated the age of the Earth to be 1.6 billion years, far older than was believed at the time. Holmes revised this estimate throughout his life, as measuring techniques improved. In 1953, an American geochemist, Clair C. Patterson, finally established the true age of the Earth at 4.55 billion years old.

Holmes also made major contributions to the theory of continental drift. This theory was proposed by German meteorologist and geologist Alfred Wegener in 1912 and states that the position of the continents on the Earth's surface has changed considerably over time. Wegener's idea was far from universally accepted, since it was not clear what would cause large continents to move across the surface of the Earth.

It was Holmes, in 1919, who suggested the mechanism: that the continents are carried by flow of the mantle on which they sit, and that the mantle is flowing because it is convecting. Warning that his ideas were "purely speculative," he suggested that rocks in the interior of the Earth would buoyantly rise toward the surface from deep within the Earth when heated by radioactivity and then sink back down as they cooled and became denser. Holmes theorized that convection currents move through the mantle the same way heated air circulates through a room, and radically reshape the Earth's surface in the process. He proposed that upward convection might lift or even rupture the crust, that lateral movement could propel the crust sideways like a conveyor belt, and that where convection turned downwards,

the buoyant continents would crumple up and form mountains. Holmes also understood the importance of convection as a mechanism for loss of heat from the Earth and of cooling its deep interior. Not until after World War II could scientists produce the hard evidence to support Holmes's fundamental concept. (To learn how this came about, read the profile of Harry Hess in Section Three.) Holmes' theories have continued to be reinforced by new data from seismologists, mineral physicists, and geochemists.

Holmes began his major work, *Principles of Physical Geology,* while standing watch against German firebombs in the laboratories of Durham University, where he was head of the Geology Department. Published in 1944 and in a substantially revised edition in 1965, shortly before Holmes's death, it is one of the most important and clearly-written books about the earth sciences. The depth and range of his thinking, which incorporated almost all aspects of physical geology, establish Holmes as a brilliant earth scientist. ❻

Section Two: How Do Scientists Explore the Inner Earth?

Fragments of periodite (light material) from the upper mantle in basalt lava (dark material). The fragments were plucked from the walls of channelways through which the magma flowed on its way to the surface.

Introduction Edmond A. Mathez

As the Earth was forming, it organized itself in layers in a process known as differentiation. The heaviest rocks sank to the center of the planet, and the lightest rocks rose to its surface. Today we observe that the Earth has four main rock layers. The deepest is the 1,300-kilometer-thick solid inner core, which is made of dense iron and nickel metal. The inner core is surrounded by a molten outer core, about 2,200 kilometers thick and made of the same material. Both the inner and outer core are extremely hot, perhaps more than 7,000°C (although we don't know their temperatures very well). Next is the mantle, which is about 2,800 kilometers thick. The mantle's composition differs greatly from that of the core—its rocks are made up of much less dense silicate minerals, which are minerals built of blocks composed of the elements silicon and oxygen. On top of the mantle is the crust, made up of rocks composed of somewhat lighter silicate minerals than in the mantle. Beneath the continents, the crust is typically about forty kilometers thick (it varies from about twenty to nearly eighty kilometers thick), but beneath the oceans it is thinner, typically only about six to eight kilometers.

Although we can't literally see the deep interior of the Earth, its slow churning, or convection, determines what part of the Earth we can see, namely, the surface, looks like. This section examines the deep interior and how scientists study it: how seismic wave data reveal the Earth's inner structures; how the Earth's convecting core generates a magnetic field; and how convection of the mantle drives the movement of rocky plates on the surface of the Earth. The latter is the process called plate tectonics, which we'll investigate further in a later section.

What is convection? It is the process by which hot material rises because it is less dense than cold material, which then sinks. Convection is a fundamental process that makes the planet dynamic. The outer core convects, the mantle convects, the oceans convect, and the atmosphere convects. As far as the solid Earth is concerned, convection is the mechanism by which heat is lost from its interior to the surface.

Convection is well-known to us. We see it when we heat a pot of water on a stove or turn on a lava lamp. In these everyday examples, convection occurs in liquids. But solid rocks, especially when they are hot and given enough time, can also flow like liquids. This is why the mantle, although solid, can convect.

Convection of the Earth's liquid outer core causes the Earth to have a magnetic field. Because it is an iron-nickel alloy, the outer core is an electrical conductor. Its convective motion generates electric currents, and these give rise to the magnetic field. This is important because the magnetic field, together with the atmosphere, shield the Earth from solar particles that would be detrimental to life.

The Earth's magnetic field has a north and a south pole. The poles wander—that is, the actual positions of the magnetic north and south poles change—at a rate that is detectable over a human life span. The wanderings of the magnetic poles reflect the somewhat irregular flow in the outer core. Even more dramatically, every several hundred thousand to several million years, the flow completely reorganizes itself in such a way as to cause the magnetic field to switch polarity: the north pole becomes the south pole, and vice versa.

We know about these magnetic reversals from the geologic record. If a rock contains magnetic minerals, the orientation of those minerals—the rock's magnetic signature—is determined by the location of the north and south poles at the time the rock cooled. By dating rocks of known magnetic polarity, geologists have developed what is termed a magnetic stratigraphy, a sort of "tree ring" record in rocks. The magnetic reversals are most dramatically visible on the ocean floor, where new crust is continually being formed along the ocean ridges. When the poles reverse, the new crust has an opposite magnetic polarity than the crust formed before the polarity change. Thus, the rocks of different magnetic polarity form in bands that are parallel to the mid-ocean ridges. The discovery of these magnetic stripes, as they have become known, was one of the major factors in the develop-ment of the theory of plate tectonics. The stripes confirmed the idea that plates were spreading apart along the mid-ocean ridges, allowing magma to rise up into the gap and freeze to form new crust.

What are some of the ways we can find out about the interior of the Earth? As described in the first essay by Robert D. van der Hilst, we deduce the internal structure mainly from seismic waves, or shock waves created by earthquakes. There are several kinds of these waves, the most important for studying the Earth's interior are P-waves, (primary, or compressional waves), and S-waves (secondary, or shear waves). P-waves travel through liquid but S-waves do not. That is how we know that the outer core is molten: there is a shadow zone where no S-waves are detected on the side of the Earth opposite an earthquake. Also, P- and S-waves travel at different speeds through rocks, depending mainly on the density of the rock. Rock density, in turn, depends on temperature and the types of minerals that compose the rock, so seismic waves give us some indication of what the Earth's interior is made. Finally, we see the boundaries of the different layers of the Earth by how the seismic

signals are affected as they pass through them. For example, although they are both made of rocks composed of silicate minerals, the mantle is denser than the crust. The boundary between the mantle and the crust, known as the Moho, is defined by seismic waves because some bounce, or are reflected, off the boundary and others change direction, or are refracted, as they pass through it. The boundary between the solid inner and liquid outer core was also discovered by diffraction of seismic waves—a discovery made in 1936 by Inge Lehmann ("the only Danish seismologist" she had joked, and certainly one of the few women seismologists of her time). Her profile is in this section.

Another group of scientists try to interpret seismic boundaries by simulating the conditions of the deep Earth in the laboratory. One of them is Elise Knittle, whose experiments are described in this section's case study. Knittle imposes very extreme temperatures and pressures on mantle rocks to see what happens. She can see, for example, how a mineral changes to a higher-pressure form to become a new mineral of the same composition—the only difference being the way the minerals' elements are packed together. This might tell us that a certain seismic boundary is due to the change in mineral structure, as opposed to a change in the chemical composition of the rock.

Another important tool for scientists is numeric modeling. Numeric modeling has become useful for understanding how the core and the mantle convect. Convection in the core is the subject of an essay by Gary A. Glatzmaier; in another essay, Peter Bunge describes convection in the mantle. Both scientists have built models on supercomputers that simulate these processes. To do that, they represent the core or the mantle as a volume, divide it into cells, and then assign each of these cells a set of properties. Next they write equations that relate the properties of one cell to those of the neighboring cells. For example, the equations relate the temperature and thus density of one cell to that of the neighboring cell, and specify how rapidly heat moves from one to the other. Then they do something to the group of cells, such as increase the heat on one side of the group, to see what happens. Since each cell influences all of its neighbors in three dimensions, as soon as there are more than a few boxes involved, a huge amount of computing power is necessary to compute how all the cells evolve in response to the disturbance. These sophisticated models have become extremely important, because they mimic nature well enough that they are testable by observations.

These computer models reveal an internal universe that is astoundingly dynamic and complex. The Earth's surface is a reflection of these massive physical and chemical processes at work in zones of inconceivable heat and pressure. This is where the real action lies: in the workings of the geodynamo—the Earth's self-perpetuating magnetic field—and in the titanic churnings of the mantle.   ❻

To investigate how scientists explore the inner Earth, I pose the following questions:

How do we know what Earth's structure is if we cannot see inside?

Robert D. van der Hilst, a Professor of Geophysics at the Massachusetts Institute of Technology, explores seismicity as a technique for examining Earth's internal structure.

What is the magnetic field and how is it generated?

Gary A. Glatzmaier, a Professor of Earth Sciences at the University of California at Santa Cruz, describes how convection in the core generates Earth's magnetic field.

What causes plate tectonics?

Peter Bunge, an Assistant Professor of Geophysics in the Department of Geosciences at Princeton University, presents computer modeling as a technique for studying mantle convection and the resulting plate tectonics.

A Kimberlite from the Mir Kimberlite pipe in Yakutia, Russia brought up this pea-sized diamond crystal, a rare sample from the Earth's upper mantle.

Global Seismic Tomography: A Snapshot of Convection in the Earth

Robert D. van der Hilst

Based on an article of that title in *GSA Today*, April 1997

Convective Flow Within the Earth's Mantle Cools the Planet

Since its formation about 4.5 billion years ago, the planet Earth has been cooling by a combination of two mechanisms. One is the relatively vigorous convection—the buoyant rise of hot material and the sinking of colder material—in the planet's interior. The second is conductive heat loss across the outermost shell of the Earth, mainly through the oceanic lithosphere—

Robert D. van der Hilst is a Professor of Geophysics at the Massachusetts Institute of Technology.

the rocky plates that form the ocean bottom, and which are much colder than the mantle below.

Sinking Slabs of Lithosphere Affect Convective Flow

Convection is the fundamental process by which our planet loses its primordial and internally generated heat. The primary force that drives convection is the downward pull of gravity on the cold, dense lithosphere, capped by oceanic crust. This downward pull causes slabs of the oceanic lithosphere to sink, or be subducted, beneath other plates, recycling these slabs back into the mantle. Rock in the mantle has to move to make space for the solid slabs of lithosphere. But how can solid rock move? It turns out that the solid rock of the mantle can flow. However, the rock that makes up the mantle is viscous, or stiff, which means that it can only flow on time scales of many millions of years. Understanding the scale and nature of convective flow within the mantle is important in order to decipher Earth's internal composition, the way it loses its heat, and the processes of differentiation that produced the planet we know today.

A long-standing goal of geophysics has been to determine the pattern of convection within the mantle. Despite years of study, several fundamental aspects of this process remain controversial. Some believe that giant convection cells extend throughout the "whole mantle" and slowly transport buoyant, hot material from the deepest part of the mantle, near the core-mantle boundary, up to the uppermost mantle, and likewise draw material from the surface of the Earth down towards the core-mantle boundary. Others believe that layers in the mantle, such as the seismic reflection at 600 kilometers depth, prevent "whole mantle" convection, causing smaller, isolated cells in the upper and lower mantle to develop. Geophysicists on both sides of this controversy

have anticipated, however, that the geometry of the subducted plates affects the pattern of convective flow within the Earth's mantle.

Scientists use seismology, the study of waves generated by earthquakes and explosions, to visualize structures that are hundreds of kilometers deep in the Earth. Because seismic waves are sensitive to both the temperature and composition of the material through which they move, interpretations of the waves' travel paths and speeds yield information about the inner Earth's density and structure. Seismic waves crisscross the interior of the Earth in all directions, and by mapping and combining their travel times and other pieces of information, scientists can arrive at a three-dimensional scan of the Earth's interior. The data are usually displayed as a series of "slices" through the planet at different depths.

The Movement of Slabs is Difficult to Infer

In the upper mantle, the location of cooler regions which are downwelling, or moving downward, can be partially inferred from the shape of current subduction zones and also, in particular, from the locations of deep earthquakes. But such unambiguous information is not available at greater depth. In the conventional subdivision of the Earth's mantle (Figure 1). the upper mantle extends from near the surface to about 660 kilometers deep, and the lower mantle from 660 to about 2,880 kilometers deep. They are separated by a well-defined seismic discontinuity, where the density of the rock changes dramatically and where deep earthquake activity in subduction zones ceases.

The Complex Boundary Between the Upper and Lower Mantle

Many studies have focused on how subducted slabs behave near this boundary. Mantle flow here is complex because the mantle minerals may change in several ways. A change in viscosity, or stiffness, may accompany changes in mineralogy that are not accompanied by changes in composition. A good example of a change in mineralogy that involves a distinct change in mineral properties without any change in composition is the transformation of graphite (the soft form of carbon that is used in pencils) to diamond (also a form of carbon but actually the hardest material known to humans). The 660 kilometer discontinuity may also mark a change in the chemical composition of the minerals, which could prohibit the transport of material between the upper and lower mantle. Detailed seismic studies suggest that beneath some island arcs the slabs of subducted lithosphere descend well into the lower mantle. In other regions, particularly beneath western Pacific island arcs, evidence exists that some slabs deflect laterally rather than sinking further downward, and spread out within the transition zone (the region

between 400 and 660 kilometers down). Computer models of mantle flow near the 660-kilometer-deep boundary that incorporate mineral phase changes, different values for viscosity, and possible changes in the composition of the rock, can reproduce the wide range of flow behavior inferred from the tomographic images without imposing a strong barrier to flow. A basic issue is the relationship between flow in the upper and the lower mantle. High-resolution seismic studies have shown that slab can descend into the lower mantle, but they have not addressed the ultimate fate of slabs and their relationship with the overall convection pattern of the lower mantle.

Figure1: Earth's internal structure delineated by seismic P- and S-waves. Seismic waves change direction and speeds with changing density at the boundaries between the major divisions of the Earth's interior: the crust, mantle, inner core and outer core.

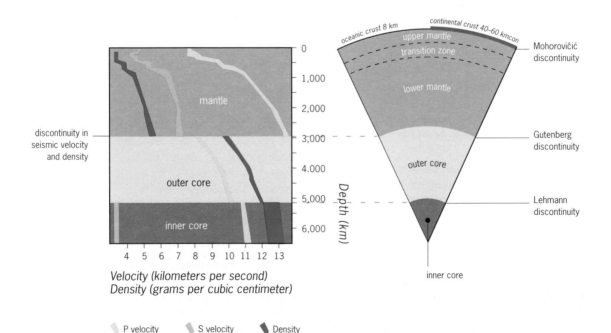

Earlier Studies of Slab Movement Disagree

Many studies confirm the existence of high-wave-speed "roots" that extend several hundred kilometers into the mantle beneath old continents, and of cooler-than-average mantle structures that are associated with subduction beneath the Pacific "Ring of Fire." While there is general agreement on structures on the order of 2,000 kilometers and more in size, results are still quite variable for smaller structures, particularly at mid-mantle depths. Until recently, seismic images of the Earth's mantle have lacked the resolution to track the flow of solid rock from the surface to the deep mantle, and thus could not be used to shed light on the controversy regarding the pattern of mantle convection. For example, a 1992 study found that higher-than-average seismic velocities within the lower mantle occur in regions with long subduction histories, and concluded that most slabs sink to the bottom of the mantle. This conclusion supports the model of "whole mantle" convection. In contrast, a 1995 study claimed a low degree of correlation between seismic structure at the top of the lower mantle and subduction history, and concluded that slabs generally remain in the upper mantle. This conclusion supports the idea of "layered mantle" convection. The fact that two fundamentally different conclusions can be reached from similar seismic studies of the mantle clearly shows that such models do not impose strict enough conditions on the geometry of flow in the Earth's mantle. Today the technique of global seismic tomography allows for improved resolution.

Two New Models Show Long, Thin Structures in the Mid-Mantle

Seismologists at different research laboratories recently constructed two new, high-resolution models that for the first time also show remarkable agreement for short wavelength structures. Both models were derived from seismic waves, but one used P-wave travel times and the other S-wave travel times. Despite the fact that they were based on two independent types of observation, both show striking high-wave-speed structures in the mid-mantle beneath the Americas and southern Eurasia that can be related to ancient subduction zones. Extremely long and narrow, the structures continue intermittently over distances exceeding 10,000 kilometers, and appear to be only several hundred kilometers thick. These structures are interpreted as subducted slabs that are continuing to sink in the mantle.

The Position of Slabs Relates to Subduction Zones

These linear high-wave-speed structures in the mid-mantle can be associated with past subduction zones in the shallow mantle. Many high-velocity zones near depths of 700 kilometers can be connected to upper mantle slabs whose existence is known from the occurrence of deep earthquakes. Even closer to the surface, there is a clear correlation between the observation of high-wave-speeds with present subduction sites. Figure 2 shows the migration of subduction zones over the past 110 million years. Note that it looks as if in the past the Farallon plate was subducting further east but in reality the North American plate moves westward to accommodate seafloor spreading in the North Atlantic and in doing so rides over the deep slab in the mantle. The two large, linear high-wave-speed structures correlate well with this history, as the deeper parts of the structures are located near more ancient subduction sites. Agreement is especially good at mid-mantle depths.

How Fast Do Slabs Sink?

The fact that long, narrow high-wave-speed structures continue from the upper mantle down as deep as 1,500 kilometers in regions of

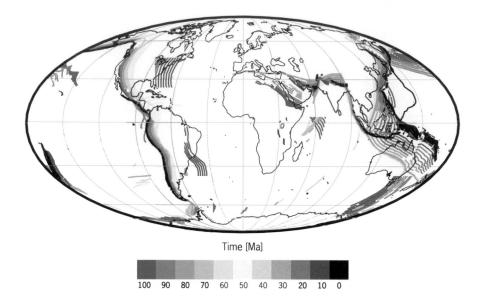

Time [Ma]

100	90	80	70	60	50	40	30	20	10	0

Figure 2: Map showing the changing positions of subduction zone locations over the last 110 million years (from model of Lithgow-Bartelloni and Richards, 1977).

ancient subduction is convincing evidence that at least in the more recent history of the Earth (within the last 200 million years) subducted slabs have sunk into the deep mantle on a large scale. Using these results and estimates of the location of the tectonic plates in the past it is possible to estimate slab-sinking rates. A significant high-wave-speed structure that extends to about 1,300 kilometers deep beneath western South America (Figure 3) serves this purpose well. It is estimated to be sinking at a rate of 0.2 to 1 centimeter/year. This is significantly less than the rate of descent in the upper mantle, which is estimated to be 5 to 10 centimeters/year. This decrease in the sinking rate implies that subduction meets with increased resistance from more viscous material as it moves deeper into the Earth.

The seismic structure of the deepest mantle is different from that of the mid-mantle. In particular, long linear high-wave-speed structures are not present, and the connection between the long wavelength structure near the base of the mantle and old subduction zones

remains tentative. More investigation of the deep mantle is needed in order to better understand this elusive part of the Earth. Thus, the studies described here suggest that the downwelling limbs of convection cells in the mantle, as defined by sinking slabs of ancient subducted lithosphere, progress at least as far as 1,700 kilometers into the mantle but may not all make it to the core-mantle boundary. The important result, however, is that downwelling limbs clearly proceed through the well-defined 600 kilometer discontinuity. This implies that convection in the mantle is likely to involve the whole mantle model rather than being confined to isolated layers.

Several Upwelling Slow-Wave-Speed Structures Also Appear

The new models described here show a pattern: most lower-mantle high-wave-speed structures can be explained by the subduction of lithospheric slabs. The nature of mantle upwelling—the upward movement of hot mantle—is far less obvious, but a few prominent, slow-wave-speed structures are apparent in our models in the deep mantle. A major deep mantle, slow-wave-speed structure appears beneath southern Africa from the

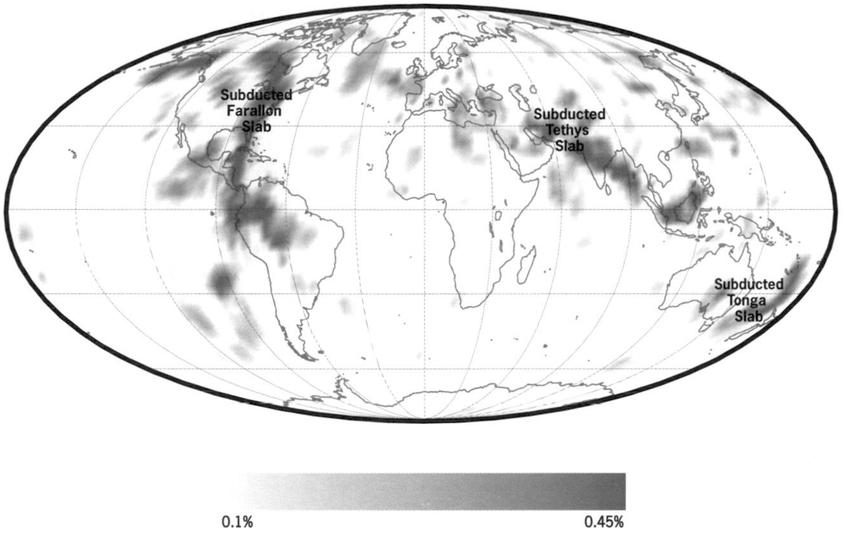

Figure 3: Map showing the high wave speed structures observed at the depth of 1,300 kilometers in the mantle. Greyscale variations indicate changes in the velocity of P-waves.

base of the mantle to a depth of nearly 1,000 kilometers. Other such structures also seen in most global models are beneath the southwestern Pacific, parts of the East Pacific Rise, the eastern Atlantic near Cape Verde, and the Atlantic near Iceland. Unlike the high-wave-speed structures, these structures are more pronounced at great depth, and tend to fade above 1,000 kilometers in depth. However, this may reflect the breakdown of the imaging technique owing to a lack of data.

Why Upwellings are Harder to Image
For several reasons, imaging upwellings is more difficult than imaging downwellings. First, upwellings are likely to occur in regions of little seismic activity and are thus not as well-characterized by seismic data, particularly in the shallow mantle. Second, our models are based on the seismic waves that arrive first, so they are biased towards high-wave-speed (faster-traveling) structures. Third, hot, thermal upwellings are believed to be narrow, perhaps no more than a couple of hundred kilometers across (although these focused upwellings do

spread out as they move into the upper regions of the mantle). This makes them more difficult to image than downwellings that form long linear features over 10,000 kilometers long. However, as imaging techniques improve and the amount of high quality data grows, we can expect to make significant progress in the imaging of hot upwellings and in constructing a more complete picture of the convection cycle.

A Unified Model of Mantle Convection is in Sight
The seismological observations discussed above are in the process of revolutionizing our views on large-scale processes in the Earth's mantle. However, reconciling the seismological and geochemical observations with insights from geodynamical models requires a concerted effort by scientists in a wide range of disciplines. (Peter Bunge's essay in this section describes in detail how geodynamicists use supercomputers to track mantle flow.) Given a high level of cross-disciplinary dialogue, we now have an excellent opportunity to construct a unified model for mantle convection and thus for understanding the major processes by which our planet is losing heat and the impact of this heat-loss process on the chemical evolution of our planet. ⊙

Convection in the Core and the Generation of the Earth's Magnetic Field

Gary A. Glatzmaier

A compass is a practical link between human beings and the Earth's magnetic field, often called the geomagnetic field. Thousands of years ago, humans recognized that iron needles made from certain iron-rich rocks (lodestones) pointed toward what seemed to be the North Star. This phenomenon was obviously useful for navigation. In the year 1600, the English

Gary A. Glatzmaier is a Professor of Earth Sciences at the University of California at Santa Cruz.

physician and physicist William Gilbert described how a compass needle changes direction around a spherical lodestone in a manner similar to the way it does over the surface of the Earth. This led Gilbert to speculate that a compass needle is actually attracted to the Earth's geographic North Pole, not to the North Star. Building on this speculation, a German mathematician named Johann Gauss demonstrated in the early 1800s that the source of the geomagnetic field is in the interior of the Earth.

What is the Magnetic Field?

A magnetic field is an invisible force field produced by moving electric charges. An atom has a magnetic field because electrons orbit its nucleus. The direction of the magnetic field is determined by the direction of the orbit. When many atoms in a material line up so that their electrons all orbit in the same direction, their magnetic fields combine to produce a large, strong magnetic field, like that around a lodestone. The north and south poles are determined by the direction of the majority of the orbiting electrons. When two lodestones, or magnets, are close to each other, there is an attractive magnetic force between the north pole of one magnet and the south pole of the other. There is also a repulsive magnetic force between the like poles.

Why Does the Earth Have a Magnetic Field?

The geomagnetic field is generated by convective motion, or flow driven by variations in density, in the Earth's iron-rich molten outer core (surrounding its solid inner core). Because the Earth is rotating, flow in the outer core is complicated and constantly changing. The flow of iron in the outer core generates electric currents, which generate magnetic fields; the fields in turn generate more currents. This process is called the geodynamo. The patterns of flow in the Earth's outer core create a magnetic field around the Earth similar to the one around a dipole magnet. This magnetic field extends throughout the entire planet and out into space, enclosing the Earth in a protective force field.

The convective motions of the inner Earth are driven by buoyancy forces. They have two main sources. The first is thermal. The temperature of the Earth increases with increasing depth. Hot material is less dense than cooler material and is therefore more buoyant. This causes it to rise. Thus, molten, iron-rich material in the core rises towards the core-mantle boundary. There it cools and becomes denser, which causes it to sink. The buoyancy forces that drive convection also have a compositional source. Iron is a relatively heavy element. We know that while the outer core is made up mostly of iron, it also contains a light element. (Some scientists think the lighter element might be sulfur, while others think it is oxygen or carbon.) As the Earth gradually cools, the molten material of the outer core is continually "freezing", or crystallizing, onto the inner core. The material that crystallizes is a dense, iron-rich alloy. As a result, the molten material left behind becomes enriched in the lighter element, becoming less dense (more buoyant). This more buoyant material then rises towards the core–mantle boundary. Thus, the compositional source of buoyancy, together with the thermal source, drives convection in the outer core. This convection generates electrical currents that in turn generate the geomagnetic field.

Why is the Magnetic Field Important for Life on Earth?

The geomagnetic field has been important historically for navigational purposes. More significantly, it extends above the surface of the planet and, together with the atmosphere, helps to shield us from dangerous cosmic radiation.

Without this force field, the radiation would reach the surface and could cause life forms to mutate.

How Does the Geomagnetic Field Behave?

The Earth's magnetic field has two poles: a magnetic north pole, and a magnetic south pole. They are closely aligned with the Earth's geographic North and South Poles, the axis of the Earth's rotation. The alignment is due to the strong influence of the Earth's rotation on the pattern of fluid motion inside the core and therefore on the generation of the geomagnetic field. However, it turns out that the alignment of both the dipole structure and the smaller component of the field that is non-dipolar are slowly and continually changing because of the chaotic nature of the convection patterns in the outer core.

In addition to these relatively minor fluctuations, every few hundred thousand years the Earth's magnetic field actually reverses, or flips over. When this occurs, the magnetic north pole becomes aligned with the geographic South Pole, and vice versa. These magnetic dipole reversals occur because of the complicated and ever-changing convective motions in the core. Each reversal is different; predicting them would be as difficult as predicting the weather for a particular day a thousand years from now. All scientists hope to do is to understand roughly how often reversals occur and how the structure of the field changes during a typical reversal.

How Do Scientists Study the Geodynamo?

Studying the cause and behavior of the Earth's magnetic field is difficult, because there is no way to directly observe the core where it is generated. The magnetic field can be studied at Earth's surface, and other physical characteristics of the core can be observed remotely. Creating a physical model in laboratory experiments using the actual materials that make up the core is practically impossible because the temperature and pressure are so extreme. Because the core is deep within the planet, and because the magnetic field changes over time scales ranging from tens of years to millions of years, the magnetic field is best studied using computer models.

This involves several steps:

1) First, geophysicists choose the laws of physics that best describe the major forces, energies, and magnetic field of the core. This leads to approximate mathematical equations that describe the most important dynamics of the Earth's core. They are approximations because only the basic physics that applies to the problem are included. The model does not attempt—and would in any case be unable—to represent the full behavior of the core. (The real world is much too complicated for even the most sophisticated computer in the world to simulate. It could never look at all the relevant variables at once, so it is necessary to constrain, or limit, them).

2) To constrain the equations, the geo-physicists assign numerical values to relatively well-known properties of the core. These properties include the core's size, its composition and material properties, and the rotation rate of the Earth.

3) Next, to constrain the equations further, they make assumptions for the less well-known properties of the core. Different assumptions give rise to different results that can then be tested against the observations. Such comparisons are a key aspect of modeling, because they can actually be used to test our assumptions and thus tell us about the lesser well-known properties. They include the rate and pattern of heat flow out of the core, and the characteristics of the boundaries between the inner and outer core and between the outer core and the mantle.

4) It turns out that because of the nature of the physics involved, the equations developed in this way don't have exact answers. So geophysicists use powerful computers to provide the best possible approximation of the real solutions. Computers are necessary because the best solutions involve an immense number of calculations—trillions of them.

Developing a Model of the Earth's Magnetic Field

The best way to go about developing a model of a complex natural process like the generation of the Earth's magnetic field is to start with the simplest possible model and see how much it explains about what we observe. Then, one step at a time, the model is made more complex to see if it captures more observations of the natural process. When it comes to modeling the core, a key but little-known

assumption relates to how heat is conducted across the core-mantle boundary. The simplest assumption is that heat is conducted uniformly across this boundary. How well does a model based on this assumption (Model 1) reproduce what we observe? Fairly well, it turns out. The Model 1 magnetic field is dominantly dipolar, and its strength is approximately Earth-like (Figure 1). The Model 1 magnetic poles line up approximately with the rotation axis of the model Earth and the alignment varies with time in an Earth-like manner. Over a simulation of nearly 500,000 years, Model 1 has shown two spontaneous magnetic reversals (one of the model reversals is shown in Figure 2A-2C). There is also a change or "drift" of the non-dipolar part of the Model 1 magnetic field with time. However, this simulated drift is more stable than the Earth's.

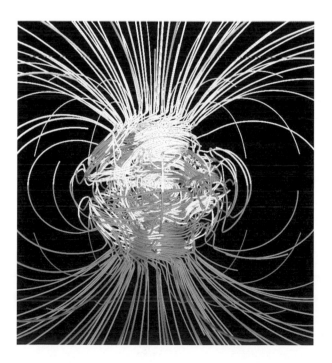

Figure1: A three-dimensional "snapshot" of the Earth's magnetic field simulated with the Glatzmaier-Roberts geodynamo model. Lines are blue where the magnetic field is directed inward and white where the magnetic field is directed outward. The rotation axis of the model Earth is vertical and through the center. The structure of the magnetic field changes at the core-mantle boundary from the intense, complicated field structure in the fluid core, where the field is generated, to the smooth, field structure outside the core.

A

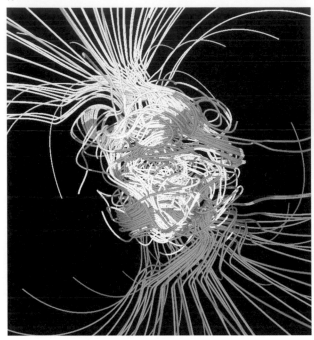

B

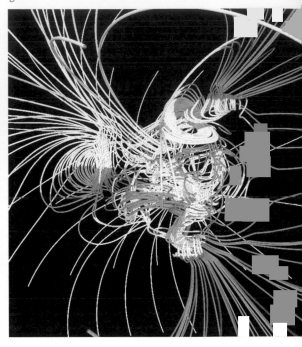

Improving the Model

One way to improve on this first model is to add layers of complexity to the magnetic field model to try to capture what is going on in the real Earth more accurately. For example, we know from an imaging technique called seismic tomography that at least in the present, heat is not conducted uniformly across the core-mantle boundary. Seismic tomography measures how seismic or energy waves, caused by earthquakes, travel through the Earth. These measurements are used to create a picture of the hot and cold regions of the mantle above the core, in much the same way that X-rays are used to image the inside of the human body. This picture makes it possible to infer the approximate pattern of heat flow out of the Earth's core. (To learn more about how seismic tomography is used to image the Earth's deep interior, read Dr. van der Hilst's essay in this section). So one way to improve the model of the geomagnetic field is to impose this observed pattern of heat flow and see if it makes a difference in the behavior of the field that is generated. This more complex

assumption of how heat flows from the core to the mantle gives rise to Model 2.

Does changing the assumption change the magnetic field simulated by Model 2? Like Model 1, the Model 2 magnetic field is dipolar and has an approximately Earth-like strength. Its magnetic poles line up approximately with the rotation axis of the model Earth. Like Model 1, the Model 2 magnetic field spontaneously reverses, and the intervals between reversals and the duration of each Model 2 reversal are similar to what has been recorded in the Earth's paleomagnetic record, while those of Model 1 are not. In addition, the non-dipolar part of the Model 2 magnetic field drifts as it does in the observed magnetic field. The simulated drift, however, is too pronounced relative to that of the real Earth. Thus, Model 1 generated drift that was too stable over time and Model 2 generated drift that was not stable enough. This tells us that the pattern of heat transport across the core-mantle boundary appears to make a difference in the behavior of the Earth's magnetic field. Together the two models may

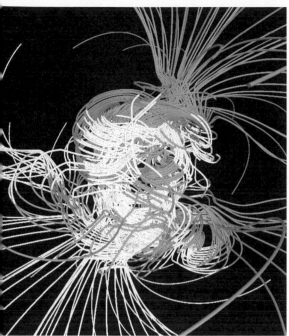

C

Figure 2:

A) Like in Figure 1, but 500 years before the middle of a magnetic field reversal.

B) At the middle of the reversal.

C) And 500 years after the middle of the reversal. Note that the blue magnetic field lines are mostly directed upwards and white lines are mostly directed downward.

be telling us that over time the pattern of heat transfer across the core-mantle boundary is something between a uniform pattern and the uneven pattern observed today with seismic tomography.

What Else Have Scientists Learned from These Computer Models?

Models 1 and 2 demonstrate that both uniform and non-uniform patterns of heat transport across the core-mantle boundary can generate reversals in the model magnetic field like those observed in the recent paleomagnetic record. In addition, the models predict that when the model magnetic field reverses, the field weakens and ceases to be strongly dipolar at the Earth's surface. During a reversal, therefore, the magnetic field no longer appears to shield the Earth from dangerous cosmic rays, but instead actually focuses them to low latitudes where most life exists on Earth. This is significant in terms of predicting how such a reversal might affect life on Earth.

In addition, both models predict that the solid inner core is spinning slightly faster than the

surface of the Earth, a phenomenon called super-rotation. This prediction was totally unanticipated, and motivated seismologists (scientists who study how seismic waves travel through the Earth) to go back to their observations to see if the prediction was valid. Current analyses of seismological data appear to support it. Thus the models have motivated scientists to see more in their observational data than they had before. This can be a valuable application of advanced computer models.

What Questions Can We Use These Models to Address?

One interesting question that scientists can explore with the magnetic field model is the impact of changes in the size of the Earth's solid inner core on the generation of the magnetic field. Remember that the Earth's core is solidifying or "freezing" over time as heat is transported from the core out through the mantle and the crust. As a result, the solid inner core is growing at the expense of the fluid outer core. The model can be used to "travel" back in time to examine the effect of a smaller solid inner core on the model magnetic field. (Early in Earth's history, the core was so hot that there was no solid inner core at all.) The model can also be used to examine the future effect of a larger solid inner core on the magnetic field. In addition, more refined models can be used to explore in detail how the next magnetic field reversal might affect life on Earth. Current models suggest that there will be an increase in the amount of cosmic radiation that reaches the surface of the Earth. More detailed predictions of the size of such an increase, as well as where on Earth the radiation might be focused, would be of great interest. ❻

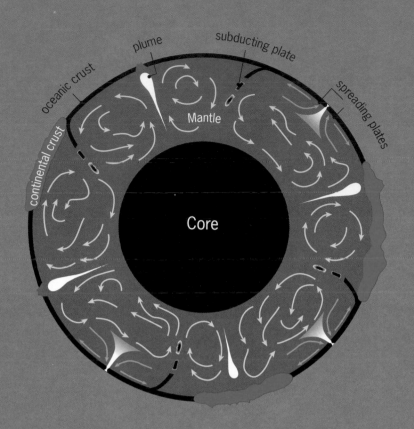

Mantle Convection

Peter Bunge

Do you believe that the Earth's interior is in constant motion?
Could I convince you that it contains currents that are tracked by
supercomputers? Well, you might say, I know about currents in
the oceans—but currents inside the Earth? After all, most of the
Earth is made up of solid rock. (There is also a big chunk of iron
in the innermost Earth, which we call the core.) Perhaps we are
both right. On a human time scale, rocks are solid, but if you
wait for a long time—millions of years—rocks slowly change
their shape. They flow.

Peter Bunge is an Assistant Professor of Geophysics in the Department of Geosciences at
Princeton University.

Over short periods of time the mantle seems rigid, but over millions of years the rocks of the mantle flow. Hot, less dense rock rises up from near the core towards the surface where it cools and eventually sinks back towards the core. This density-driven movement of material is called convection, the primary way in which heat is transferred from deep within Earth to its surface.

In my work, I use supercomputers to track the currents that churn beneath the surface of this dynamic planet.

The Theory of Continental Drift

Geologists knew this well before computers existed. The idea so fascinated Arthur Holmes, a British geologist, that he envisioned the rocks slowly deforming, or moving, deep inside the Earth, and in the process pushing continents apart on the surface of the globe. Holmes was not alone in his search to understand the motion of the continents. In fact, he was motivated by another great earth scientist, the German meteorologist Alfred Wegener, who had suggested that continents drift. At first, like others before him, Wegener was simply impressed by how the coastlines of all the continents around the Atlantic Ocean looked like they fit together. But Wegener did not stop there. Instead, he began a careful study of rocks and fossils from the edges of Africa, Australia, and North and South America. He found similar rock types along the coasts of all the continents that ring the Atlantic Ocean. He found evidence of glaciers that had originated in present-day South Africa and spread into South America and India. And he found fossils in the southern continents that must have inhabited similar ecosystems. In order to explain these similarities and connections, Wegener hypothesized that the continents must have been together at some point in the past.

What Propels the Continents?

In 1944, Holmes proposed an outrageous idea. If all continents had once been joined, they must have covered a vast region of the underlying mantle. Like a blanket, the continents would insulate this region deep inside the Earth. The region would then warm up over an expanse of geologic time. Because these warm rocks would be much lighter than the rocks on the surface, they would rise and push the overlying continents apart. This mechanism, called thermal convection, had been applied to the behavior of fluids but not to the behavior of the solid mantle. (For more about Holmes' theories, read the profile of Arthur Holmes in Section One). Convection in the mantle provided an elegant way to propel Wegener's continental drift, and the field of mantle convection was born.

How Geophysicists Use Computer Models

More than fifty years have passed, and the geophysicists who study convection in the mantle have significantly advanced Holmes' theories. They have set up mathematical equations to describe mantle convection, and they have turned to computers to help solve them. Geophysicists have long tried to use computers to better understand the Earth. Computers are useful because many complicated processes inside the Earth can be understood only by solving detailed mathematical equations over and over again. We begin by breaking up sections of the mantle into discrete areas, or cells. We track several variables in one cell, then compare these results to what is occurring in neighboring cells. These results are projected onto relatively short periods of time (maybe 100,000 years). When all these complex equations are computed, we can begin to make assumptions about the processes in individual areas over time. Based on what we know about how one area's activity relates to another, we can then make assumptions about what will happen in larger areas. Fortunately, this tedious task can now be carried out by computers. As a result, detailed computer models of the mantle now exist.

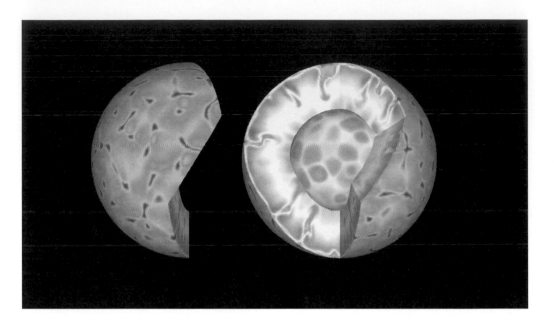

Figure 1: The results of the model show the temperature distribution for mantle convection assuming incompressible, internally heated isoviscous mantle with a Rayleigh number of Ra=10^7. The surface part is not shown, so that we can see the lighter colored upwellings and darker colored downwellings (the planform). The planform is dominated by downwelling plumes, rather than by linear zones.

These models work by laying a three-dimensional grid with millions of points over the entire mantle. The models keep track of several variables—including the velocity, temperature, and pressure at each of the grid points—by solving the appropriate mathematical equations repeatedly. After doing this many millions of times, the computer acquires enough information to form a portrait of convection in the mantle. It takes a day to two weeks to run such a simulation on a supercomputer.

The Role of Plate Tectonics

Mantle convection has turned out to be far more exciting than Holmes originally envisioned. Holmes thought only of drifting continents, but it turns out that the entire surface of the Earth is moving, a process now known as plate tectonics. The Earth's outer, rigid shell is broken up into pieces called tectonic plates. The plates are propelled by the mantle convection currents below. The continents are embedded in these plates, and are carried along as the plates move. Tectonic plates form slowly at the surface at undersea valleys, called mid-ocean spreading centers, where the mantle loses its heat to the exterior. These are places where the tectonic plates spread apart and magma, or molten rock from the mantle, wells up. The magma freezes in the gap formed by spreading, causing the plates to grow. The plates cool off as they move away from the spreading centers, and eventually they sink back into the mantle at places called subduction zones. They sink because they become colder, and therefore denser, than the mantle below. The formation of new plates and the consumption of old plates are directly coupled with mantle convection (Figure 1).

How Does Mantle Convection Work?

The nature of mantle convection can be explored in computer simulations, which sometimes pose questions as well as answer them. For example, earlier simulations predicted a complex pattern of mantle flow involving a large number of relatively small subduction zones spread across the Earth (Figure 2). In

Figure 2: For this model the viscosity of the lower mantle has been increased by a factor of 30. Linear downwelling sheets dominate the planform.

fact, we know that there are only two major subduction zone systems. One system runs from Australia to Japan along the western margin of the huge Pacific plate. The other lies off the western coasts of middle and South America, where the smaller Cocos and Nazca plates return into the mantle. Until recently, mantle convection simulations have rarely resulted in such a simple flow.

Why are Mantle Convection Cells So Wide?
Why are there so few subduction zones, especially in view of the physics of convection cells? A convection cell describes the motion in which hot material circulates upward, then sinks back down as it cools. Subduction zones are the downwelling part of a convection cell; their counterparts are the mid-oceanic ridges. In general, convection cells are as wide as they are deep, and their size decreases as the vigor of convection increases. Abundant heat, which drives convection, is generated throughout the mantle by radioactivity. Small convection cells are thus predicted for the mantle, since it is hot

and therefore vigorously convects. However, in the 3,000-kilometer-deep mantle, the cells are much wider than they are deep, averaging as much as 10,000 kilometers across. For geophysicists, this raises the question of why the mantle prefers such long convection cells rather than the smaller cells predicted based on the simple laws of physics.

Greater understanding of this puzzle has arisen through recent large numerical convection simulations based on our knowledge of the physical parameters of the mantle. The answer lies in the fact that rocks in the lower mantle more than 600 kilometers below the surface are stiffer, or more viscous, than rocks in the softer upper mantle (Figure 2). This is because pressure in the mantle increases with depth. The ever-increasing pressure causes mantle minerals to undergo a series of structural transformations so that they can be packed more closely. Such high-pressure minerals—like the mineral perovskite, which makes up more than ninety percent of the lower mantle volume and thus is probably the most abundant mineral on Earth— are stiffer and mechanically stronger than their upper-mantle counterparts. It is this contrast in viscosity between the upper and lower mantle that produces the wide convection cells observed in the mantle. The same effect is responsible for the linear nature of subduction zones.

Testing Old Theories with Seismic Tomography

Many new observations about the mantle are rendering the study of mantle convection highly exciting. Computer models are now sophisticated enough to test long-held geophysical hypotheses against these new observations. Most of this new data originates from the use of a technology called seismic tomography. Named after an imaging technique that infers the internal structure of the human body from external measurements, tomography is now being applied to the internal structure of the Earth. It is changing our understanding of the relationship between global plate tectonic motions and the upwellings and downwellings of mantle convection that drive them. (For more information about investigating the mantle with seismic tomography, read the essay by Dr. Robert D. van der Hilst in this section.) Tomographers use many types of seismic vibrations to probe the Earth. When a seismic "ray" passes through a cold, dense region or a warmer, less dense one, it speeds up or slows down correspondingly. This alters the arrival time of the ray. Thousands of such measured arrival times may be used to obtain a CAT scan image of the speed at which waves travel through different parts of the mantle. Their speeds in turn can be interpreted to reveal temperature differences between different regions. The images show that mantle flow is organized into concentrated structures. Denser structures through which waves travel at high speed correspond to regions of the Earth where cold lithospheric plates have sunk deep into the mantle at convergent plate margins, or subduction zones.

Seismic images imply a relatively simple relation between places of subduction and places where the mantle is unusually cold. In other words, the mantle temperature is low wherever old ocean floor has been subducted back into the Earth. Geophysicists can now test the link between subduction and mantle temperature with sophisticated computer models. These tests are essential because they allow us to decide quantitatively between a number of competing ideas on how the mantle works. For example, it has long been thought that no material could cross the boundary between the upper and the lower mantle, located at a depth of about 660 kilometers. That this is not the case can now be demonstrated in convection simulations, which show that old ocean floor does sink freely from the upper into the lower mantle. These same computer simulations are extremely successful in predicting the structure of the lower mantle observed by seismic tomographers.

Understanding Earth's Dynamic Past

Seismic observations and computer models of the mantle allow us to refine our understanding of the past, and give important new insight into the dynamic evolution of the Earth. Computational geodynamicists work in a highly interdisciplinary manner, drawing on the fields of seismology, mineral physics, fluid dynamics, tectonics, and computer science. An important future problem for global geodynamic models is to account more accurately for the complicated mechanical behavior of the lithosphere, the rigid outer shell of the Earth that is broken up into tectonic plates. The behavior of lithospheric plates is dominated by brittle failure along plate margins. This creates faults and generates earthquakes, behavior which is difficult to produce in the current generation of mantle models. Yet, modeling such behavior will be essential to our growing understanding of the plate tectonic convection style of Earth. ◐

Ultra-High-Pressure Experimentalist Who Studies the Deep Earth

"Because we live at the surface, we don't appreciate that the whole Earth acts as a big system," says Dr. Elise Knittle from her laboratory in the Earth Sciences Department at the University of California, Santa Cruz. "For example, we live on these big plates that are moving around slowly. Earthquakes, landforms, and volcanoes are surface manifestations of this tectonic activity, and what drives it all is heat from the interior. One main source of this heat is the Earth's core, which is almost as hot today as it was when the Earth was formed."

Cross-sections of the Earth show layers as tidy as tree rings: core, mantle, and crust. However, early in the twentieth century, geophysicists realized that in fact the boundaries between these layers are dynamic and complex. Recent evidence shows that activity at the core-mantle boundary in particular

has profound thermal and chemical implications for the way the Earth works.

In an article for the journal Science, Dr. Knittle and fellow geophysicist Raymond Jeanloz describe it as "a dynamic boundary between the rapidly convecting outer core and the slowly convecting mantle" which modulates temperature change in the deep Earth. Activity at the core-mantle boundary may also generate "hotspots"—plumes of hot mantle that rise to the surface to create volcanic island chains like the Hawaii-Emperor Seamont chain. Properties of the boundary also influence the Earth's magnetic field and the paths taken when the magnetic poles reverse.

In order to understand these important functions, scientists need to know what the core-mantle boundary is made of. It's a challenge: the layer lies almost 3,000 kilometers deep, halfway between the planet's crust and its center. At the core's edge, the temperature jumps some 1,000°C. The increase in density is greater than that between air and soil, and the pressures are tremendous—166 gigapascals, to be precise. A gigapascal is a unit of pressure (force per unit area) equaling ten to the ninth pascals, so this amount is "almost 1.4 million times the pressure at the surface," explains Knittle. Understanding the chemical interactions between the metals that make up the core and the silicate minerals that make up the mantle is essential, and that's Knittle's area of expertise. In particular, she studies the chemical and mineralogical nature of the lowermost 200–300 kilometers of the mantle, a separate layer which geophysicists refer to as D" (D double prime). It was her groundbreaking experiment that demonstrated how a chemical reaction, similar to the interaction between iron in soil and oxygen in the atmosphere that gives clay its reddish tint, is taking place between the mantle and the core.

"My research specialty is sometimes called mineral physics, because we focus on measuring the physical properties of minerals," says Knittle. "We make the link between mineralogy and geophysics. For example, a seismologist studies the speed at which waves move through Earth's interior. One of the main goals in my field is to try to understand the composition of the materials that correspond to different wave speeds. We try and take this geophysical information and translate it into real rock types in the interior." Seismic information is Knittle's evidence. It is the way geophysicists observe what they cannot see or sample.

In her laboratory, Knittle conducts physical experiments that recreate and probe the conditions of the deep Earth. The intense heat and pressure are hard to imagine, let alone simulate. It's extremely demanding, both technically and conceptually, especially since the limits of current technology make it possible to recreate those conditions only on a minute scale. Knittle's samples are flecks of mineral that can weigh just a few billionths of a gram and be less than a tenth of a millimeter across. "If I'd known my hands were this steady, I could have been a brain surgeon," she jokes.

"No one sets out to do this kind of work," Knittle continues with a smile. "Originally I was an astronomy and physics major. I got some advice that this high-pressure experimentation was an interesting way to look at planetary interiors. I wanted to understand some of the new seismic information coming out about the core-mantle boundary, and I really wanted to work on something completely new. The chemistry of the core-mantle boundary was a total unknown when I started out; no experimental work had been done."

The real drawing card for Knittle is that "it's very creative. Instruments are just stuff sitting in your lab," she points out. "I have to devise what experiments I can do to simulate this region of the Earth. What can I do technically, and what makes sense in terms of what I want to find out? Because the experiments are technically demanding—they don't work most of the time—I want to come up with an experiment that will give me the most information the couple of times it actually works. So I devised these experiments in which I combined minerals to create these mini-core-mantle boundaries." That's only part of the job. "The second challenge is how to analyze the results," explains the geophysicist. "I need to get chemical and structural information out of a sample so small it looks like a piece of pepper, and I need to figure out the best way to go about it." Some experiments are harder than others, and "any time you get a difficult experiment, there's an element of skill and one of chance," she says. Knittle's primary tool is called a diamond-anvil high-pressure cell. Small enough to be held in the palm of one hand, the device is essentially a high-tech nutcracker. At the center are two flawless, gem-cut diamonds with surfaces no wider than the heads of pins. They are ground to flat surfaces and mounted opposite each other on a hardened-steel mechanism. A fragment of material about a third of the size of the diamonds is placed between them, and force is multiplied by the mechanism by a factor of 500 to 1,000 when screws or a bolt are turned by hand. Because the area over which the force is concentrated is extremely small, the pressure—force divided by area—on the sample can be tremendous. And while it's relatively easy to generate enormous pressures for a few millionths of a second with a shock wave experiment, the diamond-anvil cell solves another problem by sustaining the pressure long enough for its effects to be studied.

The fact that the diamonds are transparent enables scientists to observe changes in the color or consistency of the minute speck of rock under pressure. It also enables them to heat the sample through the diamond window. "We heat it with an infrared laser," explains Knittle. To measure the pressure, "we put little tiny bits of ruby powder in our sample, and focus a different kind of laser, a blue one, onto these ruby chips. Under blue light, ruby fluoresces, and glows a bright red, which shifts with pressure. Then we measure the color of this fluorescence to obtain the pressure, measured in gigapascals."

"Lab work can be very tedious and time-consuming," Knittle points out. "Some new techniques don't work the first fifteen times, yet I'm sure they can work if I just do one aspect of it better. So I have a great deal of patience." She's also endlessly curious. "In science you can often get the answer that you want, but it leads you to a new question. I almost never come to the end of a project and think that we've gotten all the answers. For example, I document chemical reactions in these incredibly small samples, and then in our imagination we scale them up to what might be happening at the core-mantle boundary. That means we're going from a 100-micron sample—one-tenth of a millimeter—to something that's thousands of kilometers across."

Most of Knittle's recent work is on subduction zones, trying to figure out what happens to the slabs of oceanic crust that descend into the mantle. "Where do they go?" she wonders. Because some of the new work on the core-mantle boundary suggests that its composition is mixed, Knittle maintains that "the next logical thing would be to recreate why there are areas of melt and areas that are solid. It could be a chemical reason. For example, maybe the regions of melt are the oceanic slab pieces. I'd love to work on that problem, but I'm not exactly sure how to do it." She pauses. "But I'm thinking about it all the time." ➏

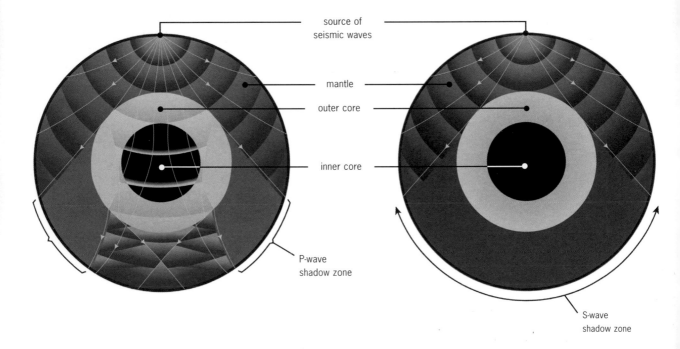

source of seismic waves

mantle

outer core

inner core

P-wave shadow zone

S-wave shadow zone

The seismic waves called P-waves pass through the core and are detected on the far side of the Earth. Indirect signals received in the P-wave shadow zone suggest there is a solid inner core deflecting some waves.

The seismic waves called S-waves do not travel through liquid. We know that the outer core is liquid because of the shadow it casts in S-waves.

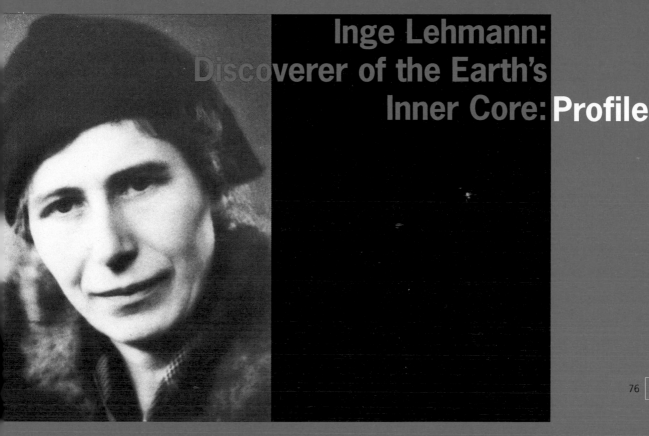

Inge Lehmann: Discoverer of the Earth's Inner Core: Profile

How can we find out what's happening deep inside the Earth? The temperatures are too hot, pressures too extreme, and distances too vast to be explored by conventional probes. So scientists rely on seismic waves—shock waves generated by earthquakes and explosions that travel through Earth and across its surface—to reveal the structure of the interior of the planet. Thousands of earthquakes occur every year, and each one provides a fleeting glimpse of the Earth's interior. Seismic signals consist of several kinds of waves. Those important for understanding the Earth's interior are P-waves, (primary, or compressional waves), and S-waves (secondary, or shear waves), which travel through solid and liquid material in different ways (see figure at left). (For more information about how seismic waves work and what they reveal, see Robert D. van der Hilst's essay in this section).

The seismograph, which detects and records the movement of seismic waves, was invented in 1880. By the end of that decade seismic stations were in place all over the world. At the time, geophysicists believed Earth to be made up of a liquid core surrounded by a solid mantle, itself surrounded by a crust, all separated by abrupt density changes in the Earth called "discontinuities."

In 1929 a large earthquake occurred near New Zealand. Danish seismologist Inge Lehmann "the only Danish seismologist," as she once referred to herself—studied the shock waves and was puzzled by what she saw. A few P-waves, which should have been deflected by the core, were in fact recorded at seismic stations. Lehmann theorized that these waves had traveled some distance into the core and then bounced off some kind of boundary. Her interpretation of this data was the foundation of a 1936 paper in which she theorized that Earth's center consisted of two parts: a solid inner core surrounded by a liquid outer core, separated by what has come to be called the Lehmann Discontinuity. Lehmann's hypothesis was confirmed in 1970 when more sensitive seismographs detected waves deflecting off this solid core.

Born in Denmark in 1888, Lehmann was a pioneer among women and scientists. Her early education was at a progressive school where boys and girls were treated exactly alike. This was a sharp contrast to the mathematical and scientific community she later encountered, about which she once protested to her nephew, Niles Groes, "You should know how many incompetent men I had to compete with—in vain." Groes recalls, "I remember Inge one Sunday in her beloved garden…with a big table filled with cardboard oatmeal boxes. In the boxes were cardboard cards with information on earthquakes…all over the world. This was

before computer processing was available, but the system was the same. With her cardboard cards and her oatmeal boxes, Inge registered the velocity of propagation of the earthquakes to all parts of the globe. By means of this information, she deduced new theories of the inner parts of the Earth."

A critical and independent thinker, Lehmann subsequently established herself as an authority on the structure of the upper mantle. She conducted extensive research in other countries, benefiting from an increased global interest in seismology for the surveillance of clandestine nuclear explosions. When Lehmann received the William Bowie medal in 1971, the highest honor of the American Geophysical Union, she was described as "the master of a black art for which no amount of computerizing is likely to be a complete substitute." Lehmann lived to be 105.   ☉

Cut away showing the four main layers of Earth: solid inner core, liquid outer core, mantle, and crust.

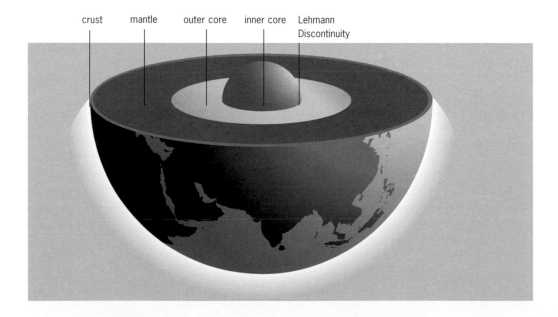

crust   mantle   outer core   inner core   Lehmann Discontinuity

Section Three: **Plate Tectonics in Action**

Detail of a cliff wall in King's Canyon in Sequoia National Forest, California, showing ductile folding of layers of chert and marl.

Introduction Edmond A. Mathez

This section is about earthquakes, volcanoes, and mountains. You might ask what mountains have to do with volcanoes and earthquakes. The answer is plate tectonics: the movement of rocky plates around the surface of the Earth. These plates make up the Earth's rigid outer shell, which is called the lithosphere. The lithosphere is a hundred or so kilometers thick, and includes the continental and oceanic crust as well as the mantle beneath them. The continents are embedded in the lithospheric plates, and move along with them. The force that drives plate tectonics—that causes the plates to move—is convection in the mantle, which has been described in Peter Bunge's essay in Section Two. Most earthquakes mark the boundaries of the plates; volcanoes are found where the plates collide, and also mark local mantle "hotspots" over which the plates move; and mountains form where one plate is thrust beneath a continent on another plate.

The earthquakes that occur at plate boundaries are caused by the ceaseless motions of the plates relative to each other. The two earthquakes of greatest magnitude in the twentieth century occurred in subduction zones, where one plate dives beneath the other and is returned to the mantle. One was the magnitude 9.2 Great Alaska Earthquake of 1964, remarkable both because of its size and the fact that it was accompanied by enormous displacements of the Earth's surface. For example, in the area of Prince William Sound, near the quake's epicenter, the ground rose as much as ten meters and horizontal movement up to twenty meters occurred, and the city of Anchorage, 130 kilometers distant, was moved about one meter to the south. Can similar earthquakes occur elsewhere? The answer is yes. As Brian F. Atwater explains in an essay in this section, geologists are particularly

interested in (and concerned about) one such area—that of the Pacific Northwest. There the geologic record indicates that great earthquakes have occurred repeatedly, although not frequently. These earthquakes must be associated with the Cascadia subduction zone, which underlies the margin of the continent from northern California to British Columbia (and gives rise to the chain of volcanoes of the Cascade Mountain range, from British Columbia's Mount Garibaldi in the north to California's Mount Lassen in the south).

Continental transform faults, along which plates slide past each other, are also regions of high earthquake activity. One of the most active is the Anatolia fault that runs across Turkey, and another is the San Andreas fault of California. There are currently no ways to predict earthquakes on faults such as the San Andreas. The best we can do is to estimate the chance that an earthquake of a certain magnitude will occur in a specific length of time—say over thirty years—along a given length of the fault. Determining the past activity of a fault helps in the making of these estimates. This is the research that Tom Rockwell does, and how he does it is described in one of the case studies in this section.

Most volcanoes in the world occur above subduction zones. These are the volcanoes that make up the world's island arcs, such as the Aleutians. They also cap certain mountain ranges, such as those that sit atop the Cascade Range. (See Tom Sisson's studies of Mount Rainier, one of the most potentially hazardous of Cascade volcanoes, described in the next section). These volcanoes are often explosive and thus can pose significant hazards to surrounding populations. Why do some volcanoes explode? That question is addressed in the essay by Haraldur Sigurdsson. The lavas produced by these volcanoes provide important clues about how plate tectonics works and what happens in subduction zones. They also tell us something about how the continents have formed, as described in Roberta L. Rudnick's essay in Section One.

A distinctly different kind of volcano comprises most of the islands in the ocean basins. This type of volcano, which is generally not explosive, erupts basalt lava, a relatively fluid material that forms by melting a small fraction of the mantle. The Hawaiian Islands are built of such material. Why do the Hawaiian Islands exist to begin with, and what do they have to do with plate tectonics? It turns out that Hawaii sits above a "hotspot" in the mantle. The Hawaiian hotspot has been supplying magma for more than 65 million years. As the Pacific lithospheric plate passes over the hotspot, the hotspot leaves a trail of volcanoes sitting on the ocean crust as evidence of its activity. The youngest volcano in this trail is near the Big Island of Hawaii. This is the volcano of Loihi, which lies to the southeast of the island of Hawaii but has yet to breach the ocean surface. The line of volcanoes, most now submerged, extends from the Hawaiian Islands northwestward, the more distant ones being the oldest. Thus, the line records the direction and rate of motion of the Pacific plate. This is described more fully in Steven L. Goldstein's essay, including how scientists think hotspots are formed. The trail of Hawaiian hotspot volcanoes will be obvious on any globe or map that shows the ocean floor, so you might want to have one handy when you read Steve's essay. This will help you imagine how the Pacific plate has moved.

Mountain ranges are a direct result of plate tectonics. Some mountain belts, like the Andes of South America and the Cascades of North America, form where an oceanic plate subducts beneath a continental one. Other mountain belts result when two continents collide. The Himalayas were formed by the collision of India with the Asian continent, a collision that took place over tens of millions of years. When continents collide to form mountain belts, the continent on one plate is thrust beneath that on the other plate, giving rise to unusually thick continental crust.

One way geologists have been exploring this process is by constructing models in which layers of sand are used to show how crusts deform. It is not as bizarre as it may sound. Sand deforms in a short time—in minutes—the way rocks do over millions of years. Jacques Malavieille is one of the scientists who constructs sand models to understand how the structures in mountains develop, and his research is described in one of the case studies in this section.

Why are mountains so high? One reason is that the compression of the crust at subduction zones causes it to become thicker. Another is that the crust is light compared to the mantle rocks below. Yet another reason is that the lithosphere upon which the mountains sit has some strength and can support them. How do you suppose we figured all this out? The story told by Peter H. Molnar is central to understanding mountains.

As far as we know the Earth is the only planet with plate tectonics. We may think of some of its manifestations, like earthquakes and volcanoes, as destructive. But in fact plate tectonics is the reason for the continents we live on and the evolution of our beautiful planet. ☯

To explore how scientists study plate tectonics, I pose the following questions:

How are mountains built?

Peter H. Molnar, a Professor of Geology in the Department of Geological Sciences at the University of Colorado, Boulder, explains the two mechanisms for supporting mountain ranges as exemplified by the Himalayas and Tibetan plateau.

How vulnerable is the Pacific Northwest to a large magnitude earthquake?

Brian F. Atwater, a geologist with the United States Geological Survey and an Affiliate Professor in the Department of Geological Sciences at the University of Washington, presents a historical perspective of Pacific Northwest earthquakes and the preparation for future earthquakes that has been based on this knowledge.

What can volcanoes tell us about the deep mantle?

Steven L. Goldstein, an Associate Professor in the Department of Earth and Environmental Sciences at Columbia University, explains the connection between hotspots and the deep mantle as he examines Hawaii, the largest hotspot on Earth.

Why do volcanoes explode and how do scientists predict an eruption?

Haraldur Sigurdsson, a Professor in the Graduate School of Oceanography at the University of Rhode Island, presents the physical and chemical processes of an explosive eruption and the signals that volcanologists can monitor that sometimes indicate a forthcoming eruption.

A view of an alpine peak from the Glarus Valley in
eastern Switzerland.

The Structure of Mountain Ranges

Peter H. Molnar

As we look at mountains, many of us are inspired by the same awe we feel before magnificent structures of human creativity, such as the soaring arches and stained glass of a Gothic cathedral. Yet, as our eyes move across the landscape, it is easy to forget that enormous forces are required not only to build but also to support a mountain range. Each range, like each cathedral, stands on a foundation without which it would collapse. In order to appreciate both mountains and buildings, it

Peter H. Molnar is Professor of Geology in the Department of Geological Sciences at the University of Colorado, Boulder.

helps to understand the invisible mechanisms that support their evident beauty. Hence the purpose of this article: to describe the underlying structure—the tectonics if not the architecture—of mountain ranges.

Two Kinds of Support

Architects have found different ways to support buildings, and these have parallels in the structure of mountain ranges. One solution is to build on a foundation of strong, inflexible rock. Correspondingly, the Himalaya, the world's highest mountains, stand on a thick shield of strong Precambrian rock. A strong rock foundation, however, is not essential. I work in a 20-story building in Cambridge, Massachusetts that rests on pilings pounded forty meters into what was once a tidal basin along the Charles River. To some extent, the building floats on water-saturated deposits, and in that respect it is not unlike a large ship. Mountain belts can also be supported by the buoyancy of light material floating on heavier material. An example is the Tibetan plateau, north of the Himalaya, nearly all of which lies above 4,500 meters.

The Survey of India

The Himalaya and the neighboring Tibetan plateau exemplify two quite distinct mechanisms for supporting mountain ranges—which is to say that both mechanisms can operate in the same region. In fact, it was the study of this area more than 140 years ago that led to the first advances in understanding the structure of mountains. The pioneers were the surveyor George Everest, the scientifically-inclined archdeacon of Calcutta, J. H. Pratt, and the eminent mathematical physicist and Astronomer Royal of Britain, George F. Airy.

In the 1840s, Everest was directing the first topographical survey of the Indian subcontinent. His crews had two methods of measuring distances. First, they could measure short distances by the conventional surveying technique of triangulation, arriving at longer distances step by step. Second, they could determine the relative positions of two widely separated points directly, by observing the position of a reference star from both points at the same time of day. In principle, the two methods should have yielded similar results, but in practice there were large discrepancies.

Everest assumed that cumulative errors in triangulation accounted for the discrepancies, but in 1854 Pratt showed that the error lay instead with the astronomical measurement. To determine the position of a star on the celestial sphere, surveyors must know precisely the direction of the zenith (the vertical direction), which was defined using a plumb line (a length of string with a weight, or bob, tied to one end). Pratt suggested that the gravitational attraction exerted by the Himalaya and the Tibetan plateau would deflect the plumb bob to the north, and that the deflection would increase with proximity to the mountains. When Pratt tried to determine the size of the error by estimating the mass of the Himalaya and the Tibetan plateau, however, he discovered that the real difference in the gravitational deflection of the plumb bob was much smaller than his calculated discrepancy. This discovery implied that he had overestimated the mass of the mountains and their substratum: there was a lot less mass under the Himalaya and Tibet than their topography had suggested.

Crust and Mantle

At first Airy was surprised by the idea of "missing" mass, but he quickly realized that rock at the surface of the Earth could not be strong enough to support a huge mass of mountains. Airy reasoned that an additional force must be present and appealed to a concept developed nearly 2,000 years ago. Archimedes reputedly first discovered what geologists now call "isostasy," better known as Archimedes Principle. As the story goes, while taking a bath he discovered a method to determine whether his king's crown was made of gold or heavier lead. He realized that the amount of water displaced by an object depended on whether it floated or sank. For example, a ten-centimeter cube of lead and a ten-centimeter cube of ice displace different amounts of water. The lead sinks, displacing its total volume, while the ice floats, displacing only about nine-tenths of its volume. This difference underlies the concept of isostasy, for in Airy's conception the Earth's light crust floated on a heavier, but presumably fluid-like, underlying layer: the mantle. The chemical composition of the crust is well known today, and it is indeed lighter and less dense than the mantle. To put it more simply, the crust floats on the mantle as cream floats on milk. Airy believed that the density of the crust is fairly uniform but that its thickness varies. The crust should be thicker under mountains, Airy argued, than under lowlands; the visible mountains are like the tips of icebergs, and like icebergs are supported by deep, invisible roots. Pratt shared Airy's conception of a floating crust, but the two men disagreed on the mechanism underlying isostatic compensation. Pratt thought the temperature, and hence the density of the crust, should vary from place to place. Where the crust is hotter and lighter than average, it stands high and forms mountains; where it is cold and dense it forms vast lowlands.

Seismological studies over the past several decades have confirmed Airy's hypothesis that the thickness of the crust varies substantially. Continental crust is on the average between thirty and forty kilometers thick, but under mountains the thickness may increase to as much as seventy-five kilometers. The crustal

roots "compensate" for the excess mass of the mountains by displacing denser mantle rock. Conversely, the crust under the deep oceans compensates for the low density of water by being only about seven kilometers thick. The force of gravity keeps the Earth in approximate isostatic equilibrium, such that the mass of an imaginary column through the Earth is roughly the same whether its surface is a mountain range or part of an ocean.

Lithosphere and Asthenosphere

Airy's version of isostasy, however, was only somewhat correct. Early in the twentieth century, seismologists found that the mantle, like the crust, is solid rather than liquid. This discovery renders Airy's image of the crust floating on the mantle an oversimplification. In the 1930s, the Dutch geophysicist Felix A. Vening-Meinesz suggested that the crust and upper mantle should distribute support for topographic loads on the Earth's surface, so that isostatic compensation should take place on a regional rather than a local scale. Support of mountains should involve more than just the formation of crustal roots. (In the limiting case this is obvious: the crust does not poke hundreds of meters into the mantle directly beneath the Empire State Building.)

Specifically, Vening-Meinesz proposed that a large load such as a mountain range deflects the Earth's strong outer layer, called the lithosphere, which commonly includes not only the crust but also the uppermost part of the mantle. The lithosphere overlies a ductile, more easily deformed layer called the asthenosphere. Under a mountain range the lithosphere bends downward, thereby distributing the weight of the range over a broad region. This bending creates a trough parallel to the range. The excess mass of the mountains is compensated in part by a mass deficit in the trough and not just by a deficit directly under the range.

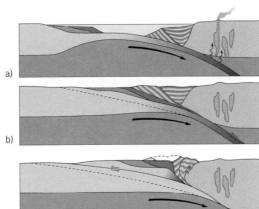

a)

b)

c)

d)

Figure 1: Himalayan mountain building. a) Subduction of the Indian Plate beneath Asia gradually closes the ocean basin between India and Tibet. b) Collision of India and Tibet and initiation of the Main central thrust fault. Sediments from the former ocean basin are forced upwards in the beginning stages of mountain building. c) Continued uplift of the Himalaya Mountains and the Tibetan Plateau. Initiation of the Main boundary fault. Erosion of former accretionary wedge sediments. d) Continued uplift and thrust fault motion on Main central and Main boundary faults.

We now know that the lithosphere consists of twenty or so separate plates. Their movements over the asthenosphere account for the formation of ocean basins and mountain ranges, and for other phenomena known collectively as plate tectonics. A lithospheric plate is somewhat like a wooden table: the table moves rigidly when it is pushed across the floor, but may sag in the middle when a heavy load is placed on it. Like the crust, the lithosphere varies widely in thickness, from negligibly thin to more than 150 kilometers.

A thick plate bends less under the weight of a mountain range than a thin one. Consequently, all else being equal, a range should stand higher on a thick plate than it would on a thin one. High mountains can nonetheless exist on a thin plate if they are supported in the way Airy envisioned: by deep crustal roots. The isostatic mechanisms proposed by Airy and Vening-Meinesz are not mutually exclusive: mountain ranges can be supported by either, or by a combination of the two; the relative importance of these mechanisms varies from range to range.

The Himalaya and Tibet

Some 70 million years ago, India and the rock that now make up the Himalaya were about 8,000 kilometers south of their present positions, drifting northward from Antarctica toward Asia on a large plate consisting primarily of oceanic lithosphere. At that time, southern Tibet lay on the south coast of Asia, about 2,000 kilometers south of its present location. As the Indian and Eurasian plates collided, the oceanic lithosphere north of the Indian landmass was bent down and thrust under Tibet (Figures 1a and 1b).

Sometime between 55 and 40 million years ago, the Indian landmass itself met the south coast of Asia (Figure 1c). The conveyor belt began to jam: the speed of the Indian plate slowed from between ten and twenty centimeters per year to about five. As India plunged under Tibet, a northward-dipping fault tore through the northern edge of the subcontinent. A slice of continental shelf and deep crust above the fault plane was, in effect, shaved off the oncoming subcontinent and thrust atop it. This off-scraping of Indian crust repeated itself at least once more. The eroded remnants of slices of ancient Indian crust form the bulk of the Himalayan range (Figure 1d).

The heavy weight of the Himalaya bends the Indian plate downward south of the range. Considering the great weight of the Himalaya, the resulting sediment-filled trough is not very deep. Because it is particularly thick, the Indian plate does not bend much. The strength of the Indian plate is a major reason that the Himalayan peaks are so high. Although the crust thickens northward beneath the Himalaya, it is much thinner than Airy's concept of isostasy would predict. The Himalaya are a good illustration of regional compensation through bending of the lithosphere as proposed by Vening-Meinesz.

In contrast, the Tibetan plateau does fit Airy's conception. The plateau extends north of the mountains for hundreds of kilometers, and in only a few valleys near the edges does its altitude drop below 4,500 meters. The underlying crust is believed to be between about sixty-five and seventy kilometers thick— thicker than the crust under the Himalayan peaks. The weight of the high plateau is compensated primarily by the buoyancy of its deep crustal root, as Airy proposed so many years ago.

Other Ranges

The Alps formed in much the same way as the Himalaya, by the stacking of slices of crustal material sheared off the southern edge of Europe and overthrust northward onto the European plate after it collided with the African plate. Less than half as thick as the Indian plate, the European plate supports a lower range than the Himalaya; the Alps reach only half the height of the Himalaya.

The Rocky Mountains in Canada also rest on a downward-flexed lithospheric plate. A gentle westward dip in the Precambrian Canadian shield beneath the mountains indicates that they are compensated regionally. Even the Hawaiian Islands depend on a strong lithosphere for their support. Their weight bends the Pacific plate downward by a few hundred meters, creating a "moat" in the seafloor around the islands.

In contrast, the Andes, the highest mountains in the Western Hemisphere, formed in a manner more similar to the formation of the Tibetan Plateau. Their weight seems to be supported by a buoyant crustal root as much as seventy kilometers deep. The Andean crust has been the focus of debate about how the crust thickens. Two different processes can occur. First, crust can be thickened by the addition of molten rock (magma) welling up from the mantle. Such rock is less dense than the mantle. Second, crust becomes thicker where its edges are pushed together by horizontal forces. Both mechanisms seem to occur side by side in the Andes.

The Mountains Are Falling

The buoyancy of the crustal root supports the weight of the Andes, but the horizontal forces that created the root also provide necessary support to the range. They buttress the Andes and prevent the range from spreading and collapsing, much the way the flying buttresses of a Gothic cathedral counter the outward forces of the heavy vaulted ceiling on the cathedral walls. Ironically, evidence for this support comes in part from the observation that the buttresses are beginning to fail. While the crust on the sides of the range is pushed together, some regions of the high Andes are spreading apart, as if the buttressing forces applied to the flanks of the Andes are not strong enough to maintain the high range.

If the Andes do collapse, they will not be the first mountain belt to undergo such a demise. Signs of crustal expansion and collapse are plentiful in the Basin and Range province west of the Rockies. There, between 80 and 50 million years ago when the Rockies were being built, a high mountain range, comparable to the Andes, once dominated the western U.S. Tibet, too, is collapsing. Although the pressure applied by India's northward motion seems to prevent Tibet from extending its north-south dimension, the high plateau has no similar buttress on its eastern flank. Accordingly, Tibet is slowly collapsing to the East.

The Andes and Tibet are particularly susceptible to collapse, in part, because they are supported mainly by deep crustal roots. The strength of the root decreases rapidly with increasing temperature and hence with increasing depth. Moreover, at a given temperature and hence depth, crustal rock is weaker than mantle rock. Therefore, thick crust tends to be weak, like ripe Camembert cheese. The Himalaya, the Alps, and the Rockies, on the other hand, are supported primarily by strong, thick lithosphere consisting of relatively cold crust and mantle, more like feta. Although these ranges were formed by horizontal forces, they do not require horizontal buttressing in order to remain standing.

Dynamics

While it is sensible to compare some mountain ranges to pressure gauges and others to loads on elastic plates, these analogies are oversimplifications. What is missing is a consideration of dynamic processes, of the forces that drive continents together, shorten the crust, cause huge terrains to be thrust onto the edges of strong plates, and drive flow in the underlying mantle. Plate motions are widely thought to be the surface manifestations of a convective circulation that extends deep into the mantle, but the overall picture of the circulation, particularly beneath mountain belts, is not well known. (For a detailed picture of mantle convection, read Peter Bunge's essay in Section Two).

Nevertheless, some underlying dynamic processes are apparent. It seems clear, for example, that Hawaii lies above a region of the asthenosphere where hot mantle material wells upward, and that upward movement elevates by several hundred meters not only the seafloor surrounding Hawaii, but the island themselves. (Read Steven L. Goldstein's essay on the hotspot origin of the Hawaiian Islands in this section). Some of the material erupts at the volcanoes on the islands, but the upwelling column is much broader than the islands themselves. It accounts for the broad swell in the seafloor around Hawaii. In contrast, one might expect to find a downwelling of relatively cold material under other mountain ranges, where slabs of lithosphere may be plunging into the asthenosphere.

Measuring Gravity

How can one study the dynamics of the mantle to determine if dense, sinking material is present under mountain ranges? One method is to measure variations in the Earth's gravity field, which should be slightly stronger above regions of dense material. Unfortunately, such differences in density are extremely small, and in mountainous regions they are masked by much larger differences caused by the topography itself. The solution is to measure gravity with satellites. With improved tracking, or with new satellite-borne instruments that directly measure lateral variations in gravity, it should eventually be possible to detect the smaller anomalies caused by density variations under mountain ranges. ❻

Averting Earthquake Surprises in the Pacific Northwest

Brian F. Atwater

Modified from U.S. Geological Survey Fact Sheet 111-95, by Brian F. Atwater, Thomas S. Yelin, Craig S. Weaver, and James W. Hendley II, of the United States Geological Survey.

How Vulnerable is the Pacific Northwest?

Until the mid-1980s, scientists and engineers believed that the threat from earthquakes in Cascadia—southern British Columbia, Washington, Oregon, and northern California—was limited to quakes not much larger than magnitude 7 on scales that seismologists use to measure earthquake size. More recently,

Brian F. Atwater is a geologist with the United States Geological Survey.

A dead forest in coastal Washington records a great earthquake that occurred in January 1700, a century before Lewis and Clark explored the mouth of the nearby Columbia River.

however, earth scientists have discovered strong evidence that great earthquakes (magnitude 8 to 9 on the Richter scale) have repeatedly struck the Cascadia region in the past several thousand years, and are likely to occur again. These shocks would be extremely large, for an earthquake of magnitude 8 would release about thirty times as much energy as an earthquake of magnitude 7, and a quake of magnitude 9 would be yet another thirty times larger. They would approach the size of the largest earthquakes ever measured—a Chilean earthquake of magnitude 9.5 in 1960, and an Alaskan earthquake of magnitude 9.2 in 1964.

The Last "Big One"
The most recent great earthquake struck the Pacific Northwest in A.D. 1700. It probably caused the ground to shake for several minutes along hundreds of kilometers of Cascadia coast and inland as well. It also set off a tsunami—a series of ocean waves caused by earthquakes or landslides on the seafloor—that deposited sand and washed over Native American fishing camps on the Cascadia coast. This same tsunami crossed the Pacific Ocean and caused damage in Japan. From the tsunami's documented arrival time in Japan, the time of the earthquake at Cascadia has been estimated as close to 9 p.m. local time on January 26, 1700.

Why the Pacific Northwest is At Risk
Had such an earthquake occurred earlier in this century, Northwesterners would have been completely taken by surprise. Now, warned by earth scientists of the likelihood of a recurrence, residents of the Pacific Northwest have taken new steps to reduce their vulnerability to future earthquake losses. Efforts to meet the threat of earthquakes in this region began decades ago, when earth scientists first recognized that the Pacific Northwest contains

a boundary between two of the tectonic plates that make up the Earth's surface. This boundary—which runs along the Pacific coast between southern British Columbia and northern California—is called the Cascadia subduction zone, and is the largest active fault in North America outside Alaska. Most great earthquakes elsewhere around the Pacific Rim occur on subduction zones. The greatest of these, in Alaska and Chile during the early 1960s, bowed nearby coastlines downward by as much as two meters, allowing sea water to flood coastlines and to kill many square kilometers of forest. Remains of the killed forests now provide geologic evidence that these earthquakes happened.

Detection of Ancient Quakes
Similar but much older remains of flooded forests are present at bays and river mouths of coastal British Columbia, Washington, Oregon, and northern California. Associated with them are sheets of sand deposited by tsunamis, and sand-filled cracks caused by strong earthquake shaking. From this evidence, scientists concluded that large earthquakes—magnitude 8 or larger—had struck the Pacific Northwest repeatedly in the past several thousand years.

The most recent of these earthquakes was the one on January 26, 1700. This date was proposed in 1996 by Japanese scientists, who identified Cascadia as the likely source of a tsunami that struck Japan but was not shortly preceded by a large earthquake in Japan. The Japanese scientists knew from American geologists that a great Cascadia earthquake had occurred between 1690 and 1715. This range was based on radiocarbon dating of earthquake-killed trees and herbs in Washington, Oregon, and California. After the Japanese scientists proposed the 1700 date, American scientists returned to some of the earthquake-killed trees. Using thin and thick rings like bar codes, they assigned dates to individual rings in earthquake-killed trees in Washington. This tree ring dating narrowed the time of the earthquake to the months between August 1699 and May 1700. Findings in the United States and Japan thus combine to give the 1700 Cascadia earthquake a place in written history, even though it predates the Pacific Northwest's earliest documents by almost a century.

Northwesterners Respond to the Risk
While watching the 1989 World Series, many Americans saw a magnitude 7.1 earthquake strike northern California. Some North-westerners were disturbed by the scenes of earthquake damage that followed and wondered if a similar disaster could strike them. By this time, many engineers and public officials had been informed about the great-earthquake threat in the Pacific Northwest. Soon after, partly in response to this threat, engineers and public officials began to revise the Uniform Building Code for Oregon and Washington.

Revising the Uniform Building Code
Earthquakes can't be prevented, but people can take measures to minimize the damage they cause. Such measures are part of the Uniform Building Code. The code contains nationwide standards for designing earthquake-resistant

structures, and these standards vary in relation to the expected level of hazard. The code defines six levels of earthquake-shaking hazard, as defined on a map that is part of the Uniform Building Code. In the early 1990s, engineers and public officials redrew this map for the Pacific Northwest. Before 1994, the code placed the Puget Sound area of Washington, including Seattle, in a zone with the second highest hazard level, and most of the rest of Oregon and Washington was placed in a zone with the next lower hazard level. The 1994 edition of the Uniform Building Code, however, extends the higher level hazard zone to include all parts of Oregon and Washington that are near potential sources of great earthquakes (Figure 1).

This revision of the Uniform Building Code is an important first step toward meeting the great-earthquake threat in the Pacific Northwest. In the areas upgraded to the second highest level of earthquake-shaking hazard, new buildings are now designed to resist earthquake forces fifty percent stronger than under the old code. The cost of these new requirements rarely exceeds a few percent of the total cost for a new building.

Figure 1: Revison of earthquake hazard zones for the Pacific Northwest.

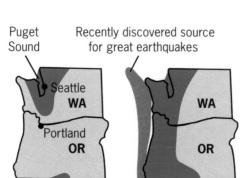

Puget Sound

Recently discovered source for great earthquakes

Seattle
WA

Portland
OR

1988

WA

OR

1994

Level of seismic hazard assigned by Uniform Building Code

High Hazard

Moderate hazard

$ millions for design and construction, 1988-1994

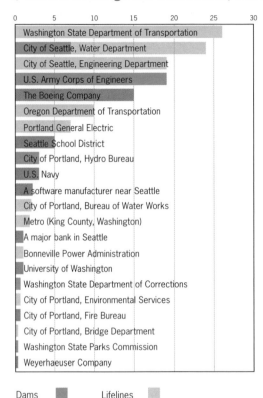

Figure 2: Outlay of funds to improve infrastructure in the Pacific Northwest.

- Washington State Department of Transportation
- City of Seattle, Water Department
- City of Seattle, Engineering Department
- U.S. Army Corps of Engineers
- The Boeing Company
- Oregon Department of Transportation
- Portland General Electric
- Seattle School District
- City of Portland, Hydro Bureau
- U.S. Navy
- A software manufacturer near Seattle
- City of Portland, Bureau of Water Works
- Metro (King County, Washington)
- A major bank in Seattle
- Bonneville Power Administration
- University of Washington
- Washington State Department of Corrections
- City of Portland, Environmental Services
- City of Portland, Fire Bureau
- City of Portland, Bridge Department
- Washington State Parks Commission
- Weyerhaeuser Company

Dams Lifelines Bridges Buildings

Protecting the Infrastructure

Discoveries about great earthquakes also helped convince public officials and corporate executives to strengthen some of the existing dams, bridges, water systems, schools, factories, and lifelines (electrical, gas, and water lines) in the Pacific Northwest. More than $130 million has been invested in such seismic upgrades since 1988 (Figure 2).

Difficult decisions must still be made about preparing for future earthquakes in the Pacific Northwest, such as:

- Should building standards near the Pacific coast be raised even further, to the highest level of earthquake-shaking hazard in the Uniform Building Code?

- Should the zone of this highest hazard level also include much of the Puget Sound area, where a large earthquake occurred 1,000 years ago on a shallow fault that passes beneath downtown Seattle?

- Should federal and state agencies spend several hundred million dollars on further increasing the earthquake resistance of bridges, as recently proposed by state highway engineers in Oregon and Washington?

How Safe Are Other Parts of the United States?

Similarly difficult questions about preparing for future earthquakes are also being asked in other earthquake-prone states. These states include Massachusetts, New York, South Carolina, Missouri, Indiana, and Utah, as well as California and Alaska. Many of the questions cannot be answered satisfactorily without more knowledge of earthquakes from the geologic past. Deciphering the geologic past is one of the ways that earth scientists help to protect people from loss of life and property.

Hawaii and Hotspots: A Window to the Deep Mantle

Steven L. Goldstein

The Earth is constantly losing heat to space from its deep interior. One indication of this, observed by miners since antiquity, is that the deeper the mine, the warmer it gets; temperatures increase by about 30°C per kilometer. The other indication is that the Earth's surface is dotted with volcanoes, which bring up heat from the deep interior.

Steven L. Goldstein is an Associate Professor in the Department of Earth and Environmental Sciences at Columbia University.

A close-up of fresh pahoehoe, or ropey lava from Kilauea volcano, Hawaii.

This heat has two sources. About half was locked into the Earth when, along with the rest of the solar system, it formed about four-and-a-half billion years ago. The Earth has since been slowly losing this primordial heat. Heat is also constantly produced by radioactive decay, mainly of three elements: potassium, uranium, and thorium. All three are present in small amounts within the Earth: 0.02 parts per million for uranium, 0.08 for thorium, and 0.03 for the radioactive specie of potassium. Yet the heat they produce, along with the primordial heat, is the reason that the Earth is a dynamic planet.

The internal heat causes the rock within the inner Earth to move and mix constantly, the way heat under of a pot of water causes the water to mix. This heat-driven process is called convection—hot, buoyant material rises, and cool, dense material sinks—providing the driving force behind plate tectonics. (To learn more about convection in the mantle, see Peter Bunge's essay in Section Two.)

This essay focuses on hotspots, small areas on the Earth's crust where enormously hot mantle comes to the surface from deep within the Earth. A large proportion of the volcanic islands in the oceans, such as Hawaii and Tahiti, were created by hotspots, as were some continental volcanoes, such as Yellowstone. In order to appreciate how hotspots work, they must be put into geological context.

Plate Tectonics and the Structure of the Earth

The Earth's crust is its thin skin, averaging a thickness of about thirty-five kilometers under the continents and six kilometers under the oceans. The mantle extends from the base of the oceanic and continental crust to the iron core approximately 2,900 kilometers deep. We know the structure of the mantle from studying the velocities of seismic waves, shock waves from earthquakes. These velocities increase

sharply at about 670 kilometers below the surface. This marks the boundary between the upper and lower mantle. A second boundary occurs at the base of the lower mantle, at a layer approximately 250 kilometers thick, which has been called the D" (d-double prime) layer. This structure has important implications for volcanism on the Earth. (To learn more about the D" layer, see the case study about the research of Elise Knittle in Section Two.)

Plate tectonics is the system which "moves" the surface layer of the Earth. This surface, which is rigid because it's cold, is made up of lithospheric plates up to a hundred or so kilometers thick. These plates are comprised of the continental and oceanic crust along with the shallowest part of the mantle that has "frozen" to the base of the crust. These plates cover the entire surface of the Earth. Several are large, like the North American plate, which extends from the middle of the Atlantic Ocean to California's San Andreas fault; many are smaller. Each plate moves across the surface of the Earth as a unit.

Boundaries between the different plates mainly follow the mid-ocean ridge system, a globe-encircling mountain range. Magmatism at the center of these ocean ridges constantly creates new ocean crust, which is why they are referred to as mid-ocean spreading centers. Like conveyor belts, the plates move outward in opposite directions from the central axis of the ridge, perpendicular to the direction of the ridge. As its distance from the ridge axis increases, the oceanic lithosphere ages and becomes colder. As it cools and the shallow oceanic mantle "freezes" to it, the lithosphere becomes both denser and thicker. Eventually, the oceanic lithosphere becomes dense enough to sink back into the mantle in a process called subduction. The same amount of crust created at the mid-ocean ridges must be lost elsewhere, because Earth does not grow. This loss occurs at subduction zones on the ocean floor called oceanic trenches. These are long, deep chasms formed where the oceanic lithosphere buckles downward and moves back into the mantle. After subduction, most of the oceanic crust gets mixed back into the mantle.

Volcanoes and Plate Tectonics

There are three major types of volcanism, each with different chemical compositions and physical distributions. Two types are associated with plate boundaries.

1) Ocean ridge volcanism associated with seafloor spreading. The largest outpouring of magma on the Earth (about twenty cubic kilometers per year) occurs under the sea at the mid-ocean ridges. As tectonic plates spread apart, the mantle is drawn upward into the central axis of the ridge. The decrease of pressure causes the rising mantle to melt partially and form magma (the melting point of almost all materials lowers when the pressure decreases). The magma then rises to the surface and erupts. In this way, the ocean ridges bring a sample of the shallow mantle to the surface. Although volcanism has played a major role in many civilizations and ancient legends, the most abundant type of magmatism on the Earth was hidden from us until the undersea explorations of the twentieth century.

2) Subduction volcanism associated with subduction zones. A line of volcanoes forms parallel to oceanic trenches, about 50 kilometers apart and 120 kilometers above the subducting oceanic crust. As the old, cold subducting oceanic crust moves downward into the mantle, it heats up. Fluids that were introduced into the ocean crust from the seawater when it was at the surface of the Earth are expelled into the mantle above. Since the addition of fluid to a solid lowers its melting point, this causes the mantle to melt and form

magma. The Pacific "ring of fire" (e.g. volcanoes such as Mount Rainier, Mount St. Helens, Pinatubo, Fuji, Cotopaxi), as well as the volcanoes in the Mediterranean (e.g. Vesuvius, Santorini), the Caribbean (e.g. Montserrat), and Indonesia (e.g. Toba, Krakatau) are associated with plate subduction. The amount of magma formed in this environment is only a small fraction—less than five percent—of the amount formed at mid-ocean ridges, but the eruptions are often explosive and have been associated with many major disasters in human history. (Haraldor Sigurdsson's essay in this section explains how explosive eruptions occur.)

3) Hotspot volcanism. These volcanoes can occur anywhere, and are usually found in the middle of plates. Most islands in the oceans are formed by this type of volcano. Over forty hotspots have been identified. Hawaii, located approximately 5,000 kilometers from the nearest plate boundary, is the biggest. The global amount of magma formed per year at hotspots is about ten percent of the amount formed at ocean ridges. While all volcanic eruptions are dangerous, hotspot eruptions tend to be fluid rather than explosive. Lava often flows slowly enough for humans to get out of the way, or there would be no Hawaii Volcanoes National Park tourist center at the rim of the active Kilauea volcano.

Hotspot Volcanism

Hotspot volcanoes, like subduction volcanoes, often form a line of volcanic islands. Generally there is only one "hot" spot; that is, there is only one island with one active volcano at a time (With two large active volcanoes, Mauna Loa and Kilauea, Hawaii is an exception.) The active volcano is located at the end of the island chain. During his voyage on the HMS *Beagle*, Charles Darwin noted that as their distance from the active volcano increases, the islands become increasingly eroded. They also appear smaller, as if sinking into the water. He

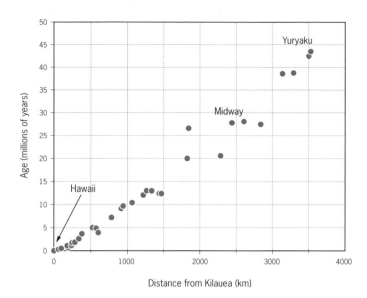

Figure 1: A plot of the age versus distance from Kilauea, the active volcano presently over the Hawaiian hotspot. The islands closest to Hawaii are youngest while those further away are oldest. Age corresponds proportionally with distance.

concluded, correctly, that the islands show an age progression (Figure 1). Eventually, the only sign of a volcano is a coral reef that forms around the undersea volcanic edifice. Finally, that too disappears. However, ocean floor maps often show a continuation of the line of volcanoes as seamounts—undersea volcano-shaped mountains—that can extend thousands of kilometers from the active volcano. The line of volcanoes and seamounts define the hotspot track.

The Hawaiian hotspot, as well as being the biggest on the Earth, has also left perhaps the best example of an age progressive hotspot track (Figure 1). The Island of Hawaii, the Big Island, is the largest Hawaiian Island and the site of current volcanic activity. On it sits Mauna Loa, the largest volcano on the Earth, rising 9,000 meters above the seafloor. A line of smaller islands, increasing in age with distance from the Big Island, extends toward the northwest. The main phase of activity on Oahu, 350 kilometers away, occurred about 3 million years ago. Kauai, 500 kilometers away, is about 5 million years old. The island-seamount chain traces a nearly straight line for about 3,500 kilometers. Midway Island, about 2,600 kilometers away and the site of the famous World War II battle, is about 28 million years

old. At the 43 million year old Yuryaku seamount, the track, now called the Emperor Seamount chain, bends sharply to the north and continues for more than another 2,000 kilometers in a roughly north-south line, to the 65 million year old Suiko Seamount on the edge of the Aleutian Trench. This means that the Hawaiian hotspot track can be traced for over 5,500 kilometers, and probably dates back farther than 65 million years. Either the volcanoes older than Suiko have been subducted back into the mantle along with the Pacific ocean crust, or they have been scraped onto the opposing plate.

How is a hotspot track made? An important clue is that the line of islands and seamounts from Hawaii to Yuryaku lies in the same direction as the movement of the Pacific Plate. In the 1960s, J. Tuzo Wilson at the University of Toronto and W. Jason Morgan at Princeton realized that the linearity of seamount chains is the result of the oceanic plate passing over a pipeline of hot, upwelling rock from the deep mantle. The rising, hot mantle plume melts and a volcano forms on the plate above. The continuously moving lithosphere carries the volcanic edifice away from the rising plume, cutting it off from the magma source. Nevertheless, the rising plume continues to produce new magma. New volcanoes are built above the hotspot, while the older ones form the chain of seamounts. About thirty-five kilometers southeast of Hawaii's Big Island, Loihi Seamount, the youngest volcano in the chain (but so new that it has yet to rise above the ocean's surface) is continuously erupting. It has grown to a height of 3,000 meters above the seafloor, and 1,000 meters below the surface. The Big Island is already moving away from the hotspot.

In effect, the Hawaiian-Emperor Seamount Chain traces the movement of the Pacific Plate over the Hawaiian hotspot over the last 65 million years. Other seamount chains on the Pacific Plate tend to be parallel to either the Hawaiian or the Emperor chain. The trend of seamount chains shows that the hotspot track traces the movement of the lithosphere over the hotspot, rather than the movement of the hotspot across the lithosphere. The "elbow" at Yuryaku Seamount shows that the direction of seafloor spreading changed about 43 million years ago.

Are the islands really sinking as distance from a hotspot increases? When the island is transported away from the hotspot, the volcanic build-up stops, but erosion continues. This erosion explains part of the decrease in elevation. However, after the island summit reaches sea level and erosion stops, the sinking continues, for two reasons. Firstly, oceanic crust cools and grows thicker as it ages, which causes it and the islands upon it to sink. Secondly, the Hawaiian plume has caused a swelling of the ocean floor around the Big Island (although there is a trough immediately around the island, as its weight pushes down on the top of the larger swell). Such swelling is characteristic of hotspots because the plume's extra heat causes the surrounding mantle to expand. The swelling is about a kilometer higher than the surrounding Pacific ocean floor, and about 1,000 kilometers in diameter. The fast

decline of elevation from the Big Island to Kauai over 500 kilometers reflects the increasing distance from the center of the swell. Although erosion also plays an important role, about 1,000 meters of the decrease in height between the Big Island and Kauai is a result of the movement of the sea floor away from the center of the Hawaiian swell.

Where do hotspots come from? Several observations are relevant to the answer.

1) A hotspot is a small, highly concentrated "point source" of volcanism that survives for long periods of time. Some, like Kerguelen in the southern Indian Ocean, have persisted for well over 100 million years. Thus, there appears to be a stable pipeline of rising hot material.

2) Through time, the volcanism occurs at the same spot. This continuous volcanism can be demonstrated by the fact that over time, two or more hotspots remain the same distance apart. Therefore, there must be a stable "root" zone in the mantle.

3) The chemical characteristics of hotspot lavas are distinct from mid-ocean ridge lavas, especially for the elements present in small quantities, called trace elements. When a rock starts to melt, some elements tend to stay in the solid residue. These are said to be compatible elements because they are compatible with the minerals that make up the solid residue. The incompatible trace elements are those that tend to go into the melt rather than remain in the solid residue. Ocean ridge basalts contain extremely low amounts of the incompatible trace elements. Geochemists interpret this to mean that the mantle that partially melts to form ocean ridge basalts had undergone previous melting episodes, because each episode further reduces the amount of the melt-favoring, incompatible elements. In contrast, hotspot basalts are much more enriched in the incompatible elements, and have been so throughout geological time. High levels of compatible trace elements are strong chemical evidence that hotspot basalts do not come from the upper mantle. Their source must be the deep mantle.

4) The products of natural radioactive decay are also powerful geochemical tracers. Like the trace elements, they show that the sources of hotspot magma are different from the upper mantle sources of ocean ridge basalts. Taking advantage of radioactive decay as nature's timekeeper, geochemists have determined that the deep mantle sources of hotspots have been isolated from the upper mantle sources of ocean ridge basalts for hundreds of millions to billions of years.

Putting these clues together, scientists think that the hotspot sources are located deeper in the Earth than the upper mantle source of ocean ridge basalts, and isolated from general convective mixing in the mantle. They remain isolated for hundreds of millions to billions of years. The two main locations where this isolation can occur are boundary layers: at the upper-lower mantle boundary at 670 kilometers, or at the core-mantle boundary. The D" layer is such a boundary.

Hotspots and the Plate Tectonic Cycle: Putting It All Together
A scenario has emerged that is generally, though not universally, favored by geologists and geophysicists (Figure 2). When oceanic crust is subducted into the mantle, some of it gets mixed back into the upper mantle by convection. A portion of it meets the upper-lower mantle boundary and pools there. Some of it passes through this boundary and gets caught up in the convection of the lower mantle. Some of this material reaches the D" layer. At these boundaries, protected from the general convection in the upper and lower mantle, the subducted crust can remain for hundreds of millions to billions of years. Eventually, the

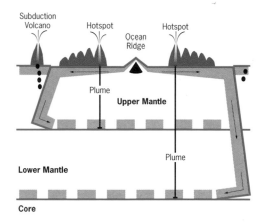

Figure 2: Diagram illustrating how subducted oceanic crust may descend to the upper-lower mantle boundary while some may go as deep as the core-mantle boundary. Some of the remelted oceanic crust may be entrained by the rising plumes that feed hotspot volcanism.

former oceanic crust is warmed by heat passing upward from the lower mantle or the core. The heat-producing radioactive elements are incompatible and are thus concentrated in the ocean crust when it forms by partial melting of the mantle. When crust is subducted into the deep mantle, its higher content of radioactive elements causes it to become slightly hotter than the surrounding mantle. This means that it also becomes less dense. Eventually a region of the boundary layer composed partly of old oceanic crust becomes lighter than the mantle above, and it begins to rise. As it rises, it establishes a conduit, or pipeline, which is rooted in the boundary layer and facilitates continued upwelling for several tens of millions of years. Eventually, the "well" dries up and the hotspot dies away. However, the boundary layers are continually replenished by oceanic crust, the source of future hotspots.

Hotspots make beautiful tropical islands and majestic volcanoes. Whereas ocean ridge basalts are our window to the shallow mantle, hotspots are our window to the deep mantle, and our only means to sample it. They provide a key reference frame for uncovering the history of seafloor spreading. They are key to understanding how the Earth works. ❻

The Hazards of Volcanoes

Haraldur Sigurdsson

Volcanic eruptions represent the most awesome and powerful
display of nature's force. The idea that the Earth may explode
under our feet and bombard us with glowing hot rock fragments
seems almost incomprehensible, but every year about fifty
volcanoes throughout the world do just that. A single eruption
can claim thousands of lives in an instant, as during the 1902
eruption of Mount Pelée on the Caribbean island of Martinique,
when a flow of hot ash and gases overwhelmed the city of
Saint Pierre and killed all but one of its 28,000 inhabitants.
More recently, a mudflow triggered by the 1985 eruption of the

Haraldur Sigurdsson is a Professor in the Graduate School of Oceanography at the
University of Rhode Island.

volcano Nevado del Ruiz in Colombia killed nearly all of the 25,000 inhabitants of the town of Armero. In addition to these devastating occurrences, hundreds of volcanic eruptions occur on the ocean floor every year, but do not generally pose a threat to humans.

One reason why volcanologists study eruptions and their effects is to develop methods to decrease the risk they pose. But before we can evaluate these volcanic hazards, we need to understand the volcanic process itself. The part that affects us is actually the last stage of a many-kilometer journey that may have taken many years.

What is a volcanic explosion? First, consider that an explosion can be either physical or chemical in origin. A volcanic explosion is physical. A physical explosion occurs when steam or another gas is held under high pressure and heat is applied, until the container—a pressure cooker, for example—suddenly breaks, releasing the steam or gas explosively and violently tearing apart the container. A volcanic explosion is similar.

In order to understand why a volcanic explosion occurs, we must first examine the liquid found within volcanoes and determine how it forms. The molten rock that erupts from volcanoes is called magma. When scientists first discovered that magmas are as hot as 800 to 1,250° C, they assumed that they originated from a molten region in the interior of the Earth. This assumption seems logical, but it turned out to be wrong. The Earth's interior, or mantle, is made of a rock called peridotite, which is hot but almost totally solid. However, when mantle peridotite experiences a decrease in pressure by rising, for example, due to the convective motion of the mantle (see Peter Bunge's essay on mantle convection in Section Two). it will begin to melt. As melting progresses, the mantle becomes rather like a sponge, consisting of solid rock containing magma.

Once magma has formed in the mantle, it rises upwards toward the crust. Magma rises because it is less dense than the surrounding rocks. The buoyant magma forces its way up through the mantle and crust, where it collects in magma chambers, typically several kilometers below the volcanoes. We have only vague ideas about what goes on inside a magma chamber—no "magmanauts" have made the journey. We cannot even send instruments or probes into the magma chamber, as the temperatures would melt even the hardest steel and the pressures would crush any instrument. However, we can examine the volcanic rock that eventually erupts from the chamber. It shows evidence that the magma in the chamber cooled and changed composition. It changes chemical composition because the crystals that form and sink in the magma chamber are of a different composition than the magma itself, so the magma is also forced to change composition. Most important as far as volcanic eruptions are concerned, the upper layers of magma also become richer in water. Under the high pressures in the chamber, this water is actually dissolved in the magma.

As more magma is added, the size of the chamber increases, and the buoyant magma begins to push upward, harder and harder, on the overlying crust. The larger the chamber, the greater the buoyancy force. To appreciate this, try the following experiment in the bathtub or next time you go swimming: hold a tennis ball in one hand and a beach ball in the other; then try to force them straight down into the water. You will have no problem with the tennis ball, but the larger ball is much more buoyant. The force you need to push it down is equal to the buoyancy force of the air in the ball. The Earth's crust is strong, but eventually the buoyancy force of magma in a large chamber equals or exceeds the strength of the overlying crust. The crust cracks, ruptures, and the eruption begins: the magma begins to move upwards.

The magma migrates up out of the chamber through cracks in the crust, called conduits, until it reaches the crater, which is the surface opening of the conduit. Most volcanoes have craters, which are closed at the bottom until an eruption occurs and opens a conduit from the bottom of the crater all the way down to the chamber. The magma may take quite a while to reach the surface. For example, the magma that erupted from Mount St. Helens in 1980 came from a chamber seven kilometers deep and moved up out of the chamber at about 0.5 meters per second, taking four hours to reach the surface.

During this journey to the surface, magma changes in several important ways. The most important change is the release of water out of the magma to form steam, a process called exsolution (Figure 1). Exsolution occurs when pressure decreases and gas separates from liquid. An everyday example of exsolution is what happens when you open a can of soda and pour its contents into a glass. The thousands of tiny bubbles of carbon dioxide gas that emerge had been in solution in the liquid until you opened the can and released the pressure. Similarly, when the magma moves upwards and pressure decreases, bubbles of steam and other gases begin to appear in the red-hot magma. But because some magma is so stiff, or viscous, the bubbles are trapped and carried along with the rising magma instead of escaping to the surface.

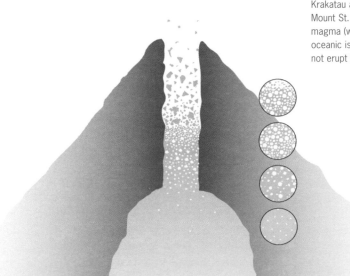

Figure 1: A volcano's explosiveness depends on the composition of the magma (molten rock) and how readily gas can escape from it. As magma rises and pressure is released, gas bubbles (mainly of water vapor and carbon dioxide) form and expand rapidly, causing explosions. Magmas with high silica content tend to erupt violently, because they are viscous. They form volcanoes like Krakatau and Tambora in Indonesia, and Mount Rainier and Mount St. Helens in Washington State. In contrast, basaltic magma (which forms the volcanoes of Hawaii and other oceanic islands) contains less silica, is more fluid, and does not erupt in giant explosions.

Another important change occurs as the magma gets nearer the surface. The pressure continues to decrease, which causes the steam bubbles to expand in size. At some point, they become so large that the magma turns into a foam. Eventually the bubbles cannot stretch the magma any further, like balloons reaching the bursting point. At this critical point the bubbles burst. They tear up the magma into small particles of molten rock, called pumice or volcanic ash, depending on the size of the particles; pumice particles are larger. Collectively, the fragments are known to volcanologists as pyroclasts (meaning "broken by fire," from the Greek words for fire, pyro, and fragments, clasts). This stage in the journey of magma—called the fragmentation level— may actually occur quite far beneath the surface, at 500 meters or so below the crater. It occurs at whatever point the strength of the bubble walls becomes less than the pressure of the gas inside the bubbles.

Meanwhile, the pressure continues to build. Now a mixture of steam, other gases, and pyroclasts moves up the conduit below the crater, but at a much higher rate than the initial 0.5 meters per second. When water exsolves from magma and forms a steam bubble at low pressure, its volume increases about 1,500 times. The exsolved steam above the fragmentation level ejects the pyroclasts forcibly out of the crater. This begins the next stage, the explosive eruption.

In the case of water-free magmas, little or no fragmentation occurs, and the relatively quiet eruption produces a lava flow. Water in the magma is therefore the key factor in establishing the nature of an eruption; its presence or absence determines whether or not the eruption will be explosive. But water in the magma does not determine whether or not an eruption will occur. The fundamental cause of an eruption is the buoyant rise of magma from the chamber.

Can We Predict a Volcanic Eruption?

Meteorologists can "see" the weather moving across the globe, using satellites and a network of weather stations to predict its behavior. Even so, they're not always right. By comparison, volcanologists have an almost impossible task. You cannot "see" the magma approach the surface. But the task is not quite impossible, thanks to the methods and instruments of geophysics, which extend scientists' sense of sight.

The magma sends out two important signals as it begins to pump up the chamber or move through the conduit towards the surface. First, in order to make room for itself, the magma must move rock, either in the chamber or in the conduit or pipe below the volcano. This space problem results in what volcanologists call "ground deformation": the surface of the volcano on the ground above the chamber begins to bulge when the magma starts to move. Sometimes the ground swells several centimeters per day. The ground deformation indicates that magma is moving and that an eruption is likely, but it doesn't tell us when it will happen.

Second, as it forces its way through brittle rock, the magma causes fractures and breakage, and each break creates a micro-earthquake, or even a full-scale earthquake. So under ideal conditions, and with sensitive seismometers installed around the volcano, we can hear the magma move out of the chamber and up towards the surface, crashing and breaking rock like the proverbial bull in a china shop.

In principle, therefore, we can monitor magma movement and predict its arrival at the surface. The problem is that only a handful of the world's volcanoes are regularly monitored in this manner, because it's very expensive to do so. Typically, the volcanoes that have recently erupted and caused much damage or loss of life are the ones that are monitored. Unfortunately, the real danger comes from long-dormant volcanoes that are not recognized as potentially dangerous. Sometimes, though rarely, mountains aren't even recognized as volcanoes until they show some activity. An even greater problem is the unpredictable appearance of a brand-new volcano as occurred in a farmer's field in Mexico in 1943. This new volcano, Paricutan, grew to over 300 meters in height in its first year. We have yet to do enough fieldwork to establish the total number of volcanoes that are potentially active (Figure 2).

NASA satellites are likely to provide us with tools to monitor all volcanoes above sea level in the future, not just the most notorious ones. Ground deformation of volcanoes can now be detected by instruments orbiting the Earth, and in future years we will probably be able to observe minute changes in the shape of an unknown volcano, predict an eruption, and provide a warning to governments and the public as magma begins to move toward the surface. 🌐

Figure 2: This map shows that most subaerial volcanoes are associated with subduction zones, places where one tectonic plate dives below another.

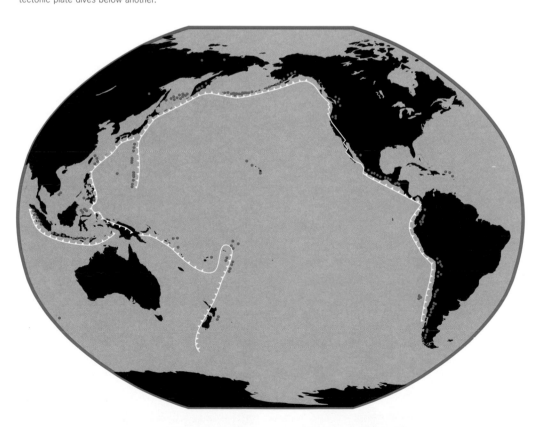

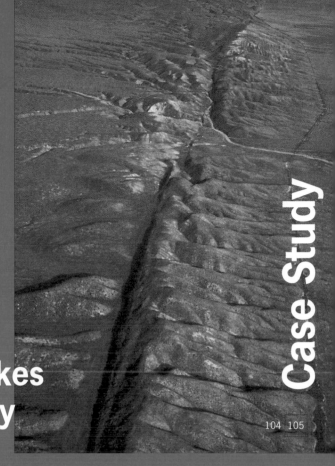

Aerial image of the San Andreas transform fault in the Southern California desert.

Forecasting Earthquakes Using Paleoseismology

"My mother first got me interested in geology," says Tom Rockwell. "We used to go out into the desert and collect rocks, and I started a rock collection when I was about six, along with my shell collection and my coin collection and all my other collections…" Now a paleoseismologist at San Diego State University, Rockwell studies the recent history of faults. A fault he defines as "a fracture in the Earth's crust across which movement occurs," and it is this movement—the friction between the two sides of the fault—that produces an earthquake. Rockwell became interested in faults and plate tectonic theory in his first geology class, and, in particular, "how faults formed in these large, worldwide systems. That eventually evolved into the work that I do now, where we go out and directly study the activity of faults and how they generate earthquakes over a period of time."

More specifically, he and his collaborators head for active fault zones all over the world, dig trenches across the faults, and examine the sedimentary layers for evidence of quakes large enough to have broken through to the ground's surface. "Our job is to try to put all the pieces of the puzzle back together, to reconstruct the earthquake history at that site," Rockwell explains. "Then we want to date the sediment as precisely as we can, so that we can then reconstruct the actual timing of past quakes." The clearest record is found in sites where fine-grained sediments have been deposited more or less continuously, right up to the present day. "If the sediments are deposited much more frequently than the quakes occur, then they should show every quake that has broken the surface at that point along the fault. It's really the sediment that provides the record," he confirms. It's a cycle. Sediments accumulate; an earthquake blasts through the sediments; more sediment is deposited across the fault, capping the layers that were deformed by the quake; another quake occurs; and on and on. "After five or six cycles, the story can get pretty messy," Rockwell acknowledges. It's a subtle story to begin with. Paleoseismologists may have to rely on delicate changes in grain size or color. Some layers are defined by nothing more than a faint line of individual grains.

The first step is to open a trench across the fault, usually with the help of a backhoe that digs a deep slot across the fault. "We'll keep the slot open with hydraulic shores—big aluminum planks held open with a piston to keep the trench from collapsing," explains Rockwell. "Then we try to cut the face of the trench perfectly flat, commonly with thin-bladed, spatula-like tools called taping knives. If it's dry and sandy, we can also use a brush."

The next step is to map out the exposure in detail, specifically its contacts. Contacts are the places where the nature of the sediment changes, for example from a sandy stream deposit to a silty flood plain deposit. The scientists place a grid of string across the exposure and document its surface as precisely as possible, using several different methods. They may photograph it and draw the contacts directly on the photograph. They may take measurements from the grid and map onto paper. Or they may skip the gridding altogether, rely on precise surveying equipment, and shoot the contacts directly into a computer which generates a picture of the exposure. According to Rockwell, "that technique is great if you don't have too many contacts."

Choosing where to dig is part technology, part research, and part intuition. To start with, Rockwell and his colleagues study aerial photography and satellite imagery for features that indicate the presence of a fault, such as

areas lifted by fault movement. "Something that you wouldn't necessary pick up on the surface stands out like a straight line on the aerial photography or satellite image," says Rockwell. He avoids old, weathered surfaces because they won't be up-to-date, and looks for fine-grained material with lots of organic matter. The sediments need to contain something like charcoal or organic peat deposits in order to be dateable by carbon-14 methods. Still, he says, "to some degree it's a shot in the dark. It depends on the geomorphology—the lay of the land—and a lot of it comes with experience." A number of new methods have been developed to study faults, including drilling and ground-penetrating radar, but according to Rockwell, "there really is no substitute for going out and digging a trench."

The only thing better than one trench is lots of them, and it's usually necessary to dig a number along any fault to characterize its behavior. "That usually also means doing a number of trenches at a single site, sometimes in three dimensions to characterize its faulting history in all directions. Ideally, we'd like to know the history every few kilometers along any given fault, but that's obviously a lot of work." Cities pose a real challenge. "Oh, it's tough," says Rockwell. "In San Diego, we did the standard look at aerial photography to locate the fault, and then we had to find a parking lot we could trench in. Sometimes in urban environments it's almost impossible, because they're almost completely built over."

Rockwell's work has taken him to more exotic places than downtown San Diego, including the Dead Sea fault zone in Israel, the northern Anatolia fault in Turkey, the Himalayan frontal thrust fault in Nepal, and the northern Gobi Desert in Mongolia. The "paleo" in paleoseismology means ancient, but in Rockwell's field it refers to any earthquake that occurred prior to the instrumental record (approximately at the turn of the twentieth century). Instrumentation is a relatively recent development, but in the Dead Sea zone a long written record exists in ancient diaries and church and mosque records. "The work in the Middle East in particular is driven by these long historical records," explains Rockwell. "This enables us to make much stronger statements about long-term fault behavior in those areas. And from this information we improve our chances of correctly forecasting future earthquakes in the U.S."

"One thing we suffer from in the U.S., in California in particular, where we have the main fault, is a short written historical record," he continues. "There are some older earthquakes that we know about from mission records, for example, but we don't know which fault produced which quakes." Paleoseismology is one way to find out. "In essence, what we're trying to do is establish a several-thousand-year record of large quakes in California and all over the western U.S. That's where I study, because that's where most of the action is."

Rockwell and his colleagues can't predict when an earthquake will occur, but they can make a statement about probability based on past history. For example, "prior to the 1989 Loma Prieta quake, the Santa Cruz mountain segment of the San Andreas fault had been forecast to have a high probability of an earthquake with the potential to do real damage. And it certainly occurred within the prescribed zone in the prescribed time frame. So you can call that a success, at least in terms of reducing hazard and making people more aware of it," Rockwell points out. "Ultimately we'd like to be able to forecast the likelihood of an earthquake in the fault in your backyard, say for the next thirty- to fifty-year period, so that you can design your building accordingly and make sure you don't build it on an active fault." On the other hand, the paleoseismologist ruefully admits, "it's a rather amazing thing, but most people do not want you to trench in their backyard and show that a fault goes through their living room. They simply don't want to know." ◐

Thomas Rockwell in Mongolia digging trenches by hand along the Bogh fault, Mongolia.

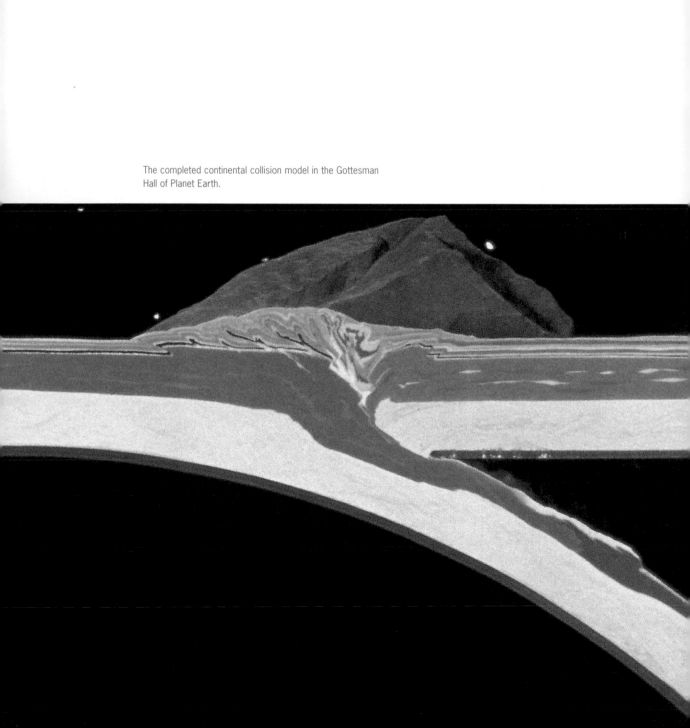

The completed continental collision model in the Gottesman
Hall of Planet Earth.

Jacques Malavieille peers through and cleans the glass before running his sand model.

A Conversation with Jacques Malavieille

Dr. Jacques Malavieille is the research director at the Université de Montpellier-2, where he runs the Geophysics and Tectonics Laboratory. He studies the formation of mountains through fieldwork and by building models in sand to understand geological processes like folding and faulting. These processes, which occur over many millennia, are replicated by the sand model within a matter of minutes. "There is absolutely no substitute for actually seeing a dynamic process happen," explains Dr. Heather Sloan, Exhibit Coordinator of the Hall of Planet Earth. "Unlike looking at static models, where how all of it works together may never dawn on you, people grasp these mechanisms intuitively when they actually see them at work." One of Malavieille's models is on display in the Hall of Planet Earth.

American Museum of Natural History: What is your field of study?

JM: I study tectonics, the formation of the Earth's crust, particularly places where continental plates move together. I study mainly mountain building, so I've done fieldwork in Tibet, in the French Alps, and the Rocky Mountains in the western United States and Canada.

AMNH: Why did you become a geologist?

JM: I was born in France, in a small city in the Massif Central, which is part of a very old mountain belt, about three hundred million years old. As a small boy I loved to look around the mountains for fossils and rocks, so maybe that's where my interest came from. Geology is an interesting science, because it combines many kinds of disciplines, physics, and mathematics, and also biology. Mountain building generally results from the interaction between two continental plates. In particular I'm studying compressional tectonics: what happens where two plates converge. To better understand what I have seen in the field, I need a simple device which reproduces the structures seen in nature—on a small scale, of course.

AMNH: How did you get the idea to make a model with sand?

JM: I wasn't the first or the only one. Many people in different countries now consider models an important tool to study geology. The first model I know of was built during the eighteenth century, by a Scottish geologist who observed that rocks were folded and deformed. So he built a small box, filled it with clay, and pushed on the layers. The structures he obtained were similar to what he had observed in the field, so he demonstrated that rocks could be deformed by a significant force. At the time, nothing was known about plate tectonics and (the forces that cause crustal) movements, so it was an important discovery.

AMNH: So generally a scientist uses a model to simulate a natural feature?

JM: Yes, that's generally the way it works. We call these analog models because we make a realistic model—an analog—of the natural object. We take care to use materials which are scaled relative to the behavior, to the time, to the size of the particular feature. For the model in the Museum I used mainly granular materials, sand and powders, because these materials are close in behavior to the rocks of the continental crust in nature at the estimated rate of deformation. For example, one centimeter of the model represented one kilometer of thickness of rocks, and one minute of running time about a hundred thousand years.

AMNH: How big is the actual model in the Museum?

JM: It is about two meters in length and maybe sixty centimeters high.

AMNH: And it represents what place on the Earth?

JM: It represents the continental crust and lithosphere, the rigid part of the Earth.

AMNH: A particular place, or the general process of mountain building?

JM: The process, because at the end of the experiment I need to have a simple structure which clearly explains what happens when two continents collide. It's a good illustration for people who don't know anything about geology.

Using computers to develop numerical models is another very popular approach. The risk here is that some people tend to use parameters which are far from the natural ones, which can have bad effects on science. You choose different parameters—the age of rocks, your measurements from the field, your observations, and so on—and run them through the computer. You must be critical of the result because it depends on what you put into the computer. Of course it's a useful and interesting tool, too.

AMNH: How is it different with your model? You also choose the parameters.

JM: The main difference is that the analog model takes into account an important fact: natural processes are not really clean. Instabilities have an important role in geology, in generating faults for example, and in the numerical model it's difficult to introduce instabilities. In the analog experiments, faults develop in materials because our natural materials mimic the behavior of rocks.

AMNH: So you have a box, you have your different grades of granular material, you have a motor. How do they come together?

JM: First, I establish special boundary conditions which imitate a subduction zone, where two continents collide and one plate goes down below the other. As you probably know, in nature, the continental crust is lighter than the oceanic lithosphere, so it cannot go down very deep. Instead it will deform, thicken, and shorten to form a mountain belt. One end of the box serves as a rigid backstop and the other is pulled at a rate that I determine. I set the motor at that rate to compress the contents of the box and simulate the actual process.

AMNH: What purposes have the models served?

JM: There are different ways to use these models. For example, through the two glass side walls, you can directly observe the dynamic of how faults and other geological features develop, the geometry of structures. Also, the model can be operated at different scales—the scale of a simple fault, for example, or the scale of an entire mountain belt—and it's easy to change a parameter. This is important because it's not always easy to decide which parameter is important. When you get an unexpected result, you have to think, why is it different? What induced the change? Probably a parameter that turned out to be more important than you thought. I discover a lot of interesting things this way.

Case Study

110 111

Detail of sand model during compression.

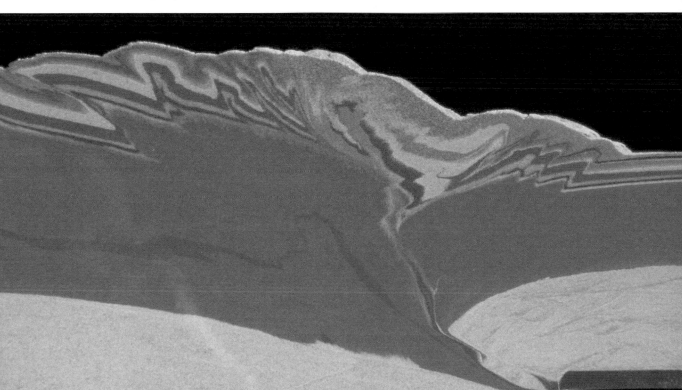

We also tape and photograph all the models as they are run, because sometimes we have to go back to old ones using new information and compare results. For example, in the Museum you have a lot of fossils, and if you discover a new one you have to compare it to the ones you've previously collected.

AMNH: So there's computer modeling, and there's analog modeling, and then there are people out there making observations, measuring mountains and so on. How do those three areas complement each other?

JM: For me it's very simple. I am first a field geologist: the most important thing is to make good observations and measurements in the field of the geological objects that you want to study. I also do physical modeling. I combine this with computer modeling. I combine approaches and compare the results with other people, even if it's not in the same university.

Geology is not an exact science like mathematics. You must be curious, you must ask other people from different disciplines for their view of the object you are studying—which is the Earth, a very, very complex subject.

AMNH: What's the one new thing that you want students to learn from seeing your physical model?

JM: From an outside view, geology seems static; we only have a snapshot of processes which may have taken thirty million years to achieve. The model establishes the relation between the geological time scale and the human one. Immediately the student can imagine that mountains are not fixed structures, that everything is moving on the Earth's surface. It shows a dynamic process. ❻

Close-up of a polished rock face that shows a fault. Note that the dark layer has been broken in two and that the two pieces have been displaced from one another.

Harry Hess (1906–1969) in his World War II Naval uniform.

Harry Hess: One of the Discoverers of Seafloor Spreading: Profile

Nothing could feel more solid than the ground under our feet. Yet the surface of the Earth is not fixed, but rather broken up like a jigsaw puzzle into enormous plates that move. This process is called plate tectonics, and it transformed the thinking of geologists. One of them, Harry Hess, was an instrumental figure in figuring out how plate tectonics worked.

Hess possessed two valuable skills: careful attention to detail and the ability to form sweeping hypotheses. This unusual combination produced groundbreaking work on a number of subjects, including the origin of ocean basins and island arcs, mountain building, and the movement of continents. The idea that the continents might have moved, or "drifted" over time can be traced back to the sixteenth century, when European cartographers compiled world maps based on the seagoing expeditions of that time. This idea was transformed into the theory of "continental drift" by German meteorologist Alfred Wegener in 1912, when he published a treatise with several lines of supporting evidence that went beyond simply matching the continents like puzzle pieces. These lines of evidence included, for example, matching geological formations and paleontological distributions from South America and Africa. Wegener's critics correctly pointed out, however, that the continents could not simply "plow" though the ocean floor as Wegener had vaguely theorized. It was Hess

who determined how oceanic mountain ranges, called mid-ocean ridges, are fundamental to the tectonic movement that results in the drift of continents.

According to his own account, Hess flunked his first course in mineralogy at Yale and was told he had no future in the field. Nevertheless he stuck with it, and was teaching geology at Princeton when World War II was declared. Already a lieutenant junior grade in the Naval Reserve, Hess was called to active duty after Pearl Harbor and was eventually to rise to the rank of Rear Admiral. He soon developed a system for estimating the daily positions of German U-boats in the North Atlantic, and requested duty aboard a decoy vessel in order to test the program. It worked.

He then served as commander of the attack transport U.S. Cape Johnson in the Pacific Ocean, taking part in major landings at Marianas, Leyte, Lingayen Gulf, and Iwo Jima. Ever the scientist, while cruising from one battle to the next, Hess kept the transport's sounding gear (which bounced sound waves off the sea-floor in order to determine the underwater relief or topography) running day and night. This led to his discovery of submerged and curiously flat-topped mountains that he named "guyots" in honor of the Swiss founder of the Princeton geology department. It also produced thousands of miles of echo-sounding surveys of the ocean floor.

The postwar period was a revolutionary one for the earth sciences. Efforts to map the ocean floor intensified, thanks in large part to the newly-created U.S. Office of Naval Research. Within a few years, a curious terrain had emerged: vast, flat plains interrupted by ridges, or more precisely, vast mountain ranges. In the Atlantic Ocean, the "ridge" is about midway between the continents on either side, and thus it became known as a mid-ocean ridge. We now know that the ocean ridge system snakes around the entire globe in a continuous chain some 80,000 kilometers long. In 1953, scientists discovered that a prominent valley, called the Great Global Rift, ran down the center of these ridges. Intrigued, Hess reexamined the data from a completely fresh, unorthodox perspective. In 1962, he proposed a groundbreaking hypothesis that proved vitally important in the development of plate tectonic theory. It addressed several geologic puzzles: If the oceans have existed for at least 4 billion years, why has so little sediment accumulated on the ocean floor? Why are fossils found in ocean sediments no more than 180 million years old? And how do the continents move?

Hess theorized that the ocean floor is at most only a few hundred million years old, significantly younger than the continents. This is how long it takes for molten rock to ooze up from volcanically active mid-ocean ridges, spread sideways to create new seafloor, and disappear back into the Earth's deep interior at the ocean trenches. This "recycling" process, later named "seafloor spreading," carries off older sediment and fossils, and moves the continents as new ocean crust spreads away from the ridges.

Supporting Wegener's theory of continental drift, Hess explained how the once-joined continents had separated into the seven that exist today. The continents don't change

dramatically or move independently, but are transported by the shifting tectonic plates on which they rest. The theory also explained Hess's puzzling guyots. They are believed to be once-active volcanoes that rose above the surface like modern-day island arcs and then were eroded to sea level. As the ocean crust spread away from the higher ocean ridges, the guyots sank below sea level, becoming completely submerged. Hess also theorized that because the continental crust was lighter, it didn't sink back into the deep earth at trenches as did the oceanic crust. Instead, it scraped rock off the descending ocean crust and piled it into mountain rages at the trenches' edge. Hess also incorporated the idea proposed by Swiss geologist Emile Argand in the 1920s that mountain belts are also created when two continents collide.

Hess's bold intuition was subsequently corroborated. Later studies showed that the age of the ocean floor increases with distance from the ridge crests, and seismic studies confirmed that the oceanic crust was indeed sinking into the trenches. His report, *History of Ocean Basins*, was formally published in 1962 and for some time was the single most referenced work in solid-earth geophysics.

Hess also contributed significantly to his university, where he became head of the Princeton geology department, and was an important member of the national scientific community. He helped design the national space program, and was one of ten members of a panel appointed to analyze rock samples brought back from the Moon by the Apollo 11 crew. He died in August, 1969, a month after Apollo 11's successful mission. A National Academy of Sciences memoir calls Hess "one of the truly remarkable earth scientists of this century." ❻

Section Four: **How Do Scientists Read the Rocks?**

Introduction Edmond A. Mathez

Geologists map the distribution of rocks; they record how rocks have been contorted over the ages; they study thin sections, which are slivers of rock through which light can pass; they determine the compositions of rocks as well as the compositions of the individual minerals that make them up; they determine rock age. In other words, geologists "read" rocks in order to deduce the geologic processes that formed them and the history they experienced. Then they piece together these histories from various sources to arrive at a history of part of Earth. In this section, we investigate how geologists approach the problem of deducing Earth's history and the processes that shape it.

The Grand Canyon is an extraordinary place to read the rocks, so the first essay in this section is by George Billingsley, who has spent many years mapping there. The Grand Canyon is an immense chasm whose walls expose an enormous span of Earth history: the rocks at the very bottom are 1.7 billion years old, while those at the top formed 260 million years ago. In between is a thick sequence of sedimentary rock layers. The story they tell is interrupted in places by unconformities—the gap left when uplift and erosion tore out a few pages of the book—but otherwise the Grand Canyon's rocks provide a continuous, visible record. How do the rock layers tell us about geologic history? To take a simple example, suppose we find a sandstone layer made up of the kind of sand found on beaches. On top of that layer is one made of mud, like the mud that accumulates in deeper water, say, at the far end of a river delta that protrudes into the ocean. On top of that is limestone, which forms in quiet and shallow seas distant from land. The rock types and sequence of layers tell us that the region was once at the shoreline and that the sea invaded, pushing the shoreline further and further inland.

(They don't tell us why this happened, however; the sea level could have risen, or the continent could have sunk.)

Some obvious rules have emerged about reading the rocks. One is the rule of superposition: that in any sequence of layers, the one at the bottom is the oldest. This rule now seems pretty obvious, but it was unknown until the early eighteenth century. Another important concept is the Uniformitarian Principle, which was first formulated in the late eighteenth century by a Scottish geologist named James Hutton, profiled in this section. The Uniformitarian Principle is important because it leads to the concept of geologic time. It states that the processes that are operating today are the same processes that operated in the past. This implies that if silts are presently accumulating in a delta at the rate of so many centimeters per hundred years, then that was the rate at which they accumulated in the past. Following this line of reasoning, geologists began to realize that a layer of rock hundreds of meters thick represents a long span of time. There are other rules, but these are two of the most important ones.

In another essay in this section, Lowell Dingus and David B. Loope set out to solve a murder mystery. The victims are a group of dinosaurs, mammals, and lizards that lived 80 million years ago in what is now the Gobi Desert. They seem to have had the misfortune of having been buried rapidly by windblown sand. It seemed odd to Dingus and his colleagues that these creatures would be just standing around while blowing sand covered them. How to solve this mystery? Again, it is by looking at the rocks.

This section also includes a case study about the work of Tom Sisson, a geologist who for the last six years has been mapping Mount Rainier.

Looming above Puget Sound in the state of Washington and the many people who live there, Mount Rainier is one of the most dangerous volcanoes in North America. The point of Sisson's work is to understand how this particular volcano works. He seeks to know, for example, the types and frequencies of eruptions that occurred in the past, in hopes of being able to tell us what to expect in the future. This is hardly what Dingus does to try to solve his murder mystery in the Gobi Desert, but both Sisson and he use the same techniques: they map, and look closely at the rocks. ☺

To explore how scientists "read," or interpret, the record of Earth processes preserved in the rock, I pose the following questions:

How was the Grand Canyon formed?

George Billingsley, a geologist with the United States Geological Survey, summarizes several theories and the erosion processes they suggest.

What can we learn about geology from dinosaur fossils found in rocks?

Lowell Dingus, a Research Associate in the Department of Vertebrate Paleontology at the American Museum of Natural History, and his colleague **David B. Loope**, a Professor in the Department of Geosciences at the University of Nebraska, analyze dinosaur fossils from Mongolia's Gobi Desert in an effort to describe the environment in which these ancient animals lived and died.

A fossil of an oviraptorid dinosaur on its nest, found in the Gobi desert.

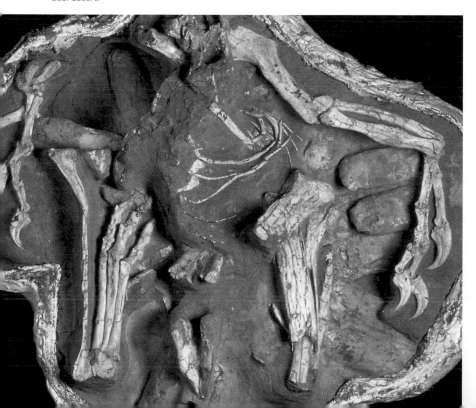

The Erosion of the Grand Canyon

George Billingsley

The Grand Canyon is a masterpiece of erosion across time. Consider just how long a million years is, and how many storms have occurred during that period, each one eroding the rocks just a little more. Although the processes of erosion may seem obvious, those that shaped the Grand Canyon are hotly debated among geomorphologists. (Geomorphology is a branch of geology that studies the general configuration of the Earth's surface and the evolution of its landforms.) The arguments primarily involve the earliest history of the Canyon, rather than the landforms produced fairly recently—within the last 5 to

George Billingsley is a geologist with the United States Geological Survey.

9 million years. How, for example, did the Colorado River become established in its present course?

An Early Theory is Contradicted by Newer Evidence

The starting point of the Grand Canyon is Lees Ferry, Arizona, and the end point is Grand Wash Cliffs, Arizona, 440 kilometers away (via the Colorado River). The canyon includes all the tributary streams and minor canyons that drain into the Colorado River between Lees Ferry and Grand Wash Cliffs. Most early geologists thought that the Colorado River had established its present course across the Colorado Plateau's relatively flat surface several tens of millions of years ago, when the Colorado Plateau slowly began to rise. (Scientists disagree about what forces caused the uplift as well as the timing of the uplift.) According to this scenario, as the plateau rose, the Colorado River simply eroded down along its present course, through buried rocks and folded structures. These structures form the higher mountain plateaus visible today, such as the Kaibab Plateau in the eastern Grand Canyon.

Clarence E. Dutton, a pioneer geologist who worked on the Colorado Plateau in the early 1900s, recognized evidence of a period of erosion that preceded the cutting of the plateau's present canyons. Dutton designated this erosional epoch as "the Great Denudation," during which the rock layers were gradually stripped away by erosion, resulting in a relatively flat landscape.

Several geologists have since found evidence that in the past the Colorado River was not a single, through-flowing stream but made up of several different, unrelated stream systems. In addition, recent geologic mapping suggests that the western part of the Grand Canyon is much older than the eastern part, because

older drainage valleys are still preserved there. To further complicate matters, the sedimentary rocks deposited against the Grand Wash Cliffs at the mouth of western Grand Canyon were not deposited by the Colorado River. Instead, those rocks were deposited by streams coming from Nevada, before the Colorado River came into existence. The evidence for this is the presence of nine-million-year-old volcanic rocks within the Grand Wash Cliffs sediments. The Colorado River then eroded into these rocks, and its gravel in turn was covered 4 to 6 million years ago by younger lavas. This means that the Colorado River must have come into existence at the mouth of the Grand Canyon between 6 and 9 million years ago.

Where Did the Action of the Colorado River Originate?

Another point of disagreement has to do with how and when the Grand Canyon got its shape. Imagine the flat plateau that stretched across the region before the Grand Canyon existed. Did the Colorado River flow down from a higher surface across this plateau, erode down into the existing landscape, and carve out the canyon in the process? Or did the Colorado River start out as a smaller river east of the canyon area, and get captured by a steeper river system from the west? According to the second scenario, the steeper river system would then have cut out the canyon as it eroded toward the smaller river. Are there other ways the Grand Canyon may have developed?

What Do The Walls of the Canyon Reveal?

The various rock formations seen in the walls of the canyon were deposited before the Grand Canyon was cut. After all, a valley has to erode through something. The oldest rocks are at the bottom of the Canyon and the youngest are at the rim, following the geological rule of superposition. Many of the older rocks deep in the canyon were once sedimentary layers that

were subsequently subjected to heat and pressure, chemically altered, and intensely deformed to form metamorphic rocks. However, most of the rocks that form the walls of the Grand Canyon are comparatively undeformed sedimentary layers, deposited upon the older rocks, which range in age from about 570 to 245 million years and were deposited in dozens of different environments. Like erosion, deposition is an ongoing process that occurs at various rates in different places: in oceans and rivers, in mud plains, and in windblown deserts. At some point in its history, the Grand Canyon encompassed each of these environments. Even younger sedimentary rocks (during the Age of Dinosaurs) once covered the entire Grand Canyon area, but these have largely been removed by early plateau erosion. These younger, red sedimentary rocks are visible to the north and east of the Grand Canyon. The youngest rocks within the Grand Canyon, which are present only in its western reaches, are volcanic and were deposited within the last several hundred thousand years.

Many Factors Shape the Canyon
Several erosional processes at work today continually fine-tune the landscape of the Grand Canyon. Most importantly, composition and hardness of the rock affect how the landscape erodes and develops, shaping buttes and mesas within the Grand Canyon. Most of the rocks in the Grand Canyon are layered sedimentary rocks of various hardness. In this arid climate, the softer siltstone and shale layers are less resistant to the forces of erosion and tend to form slopes. The harder limestone and sandstone are more resistant, and tend to form cliffs. The cliff-slope relationships between soft and hard rock layers in the Grand Canyon have produced the striking stair-step topography of the canyon walls.

What General Processes Shape Valleys?
The Grand Canyon is a valley. Valleys are so common on the Earth's surface that they go by many names—gully, draw, ravine, gulch, hollow, run, arroyo, gorge, and canyon, or for more poetic terms, vale, glen, and dale—yet we seldom bother to define them. Valleys are landforms of varying size and shape occupied by streams that are either perennial (always there) or intermittent (sometimes there). All valleys were cut by flowing water or ice. A valley takes form through three related processes: valley deepening, valley widening, and valley lengthening. Valley lengthening happens when a valley establishes its basic shape. How the Grand Canyon did so is in dispute, as discussed earlier. Valley deepening and valley widening are currently in progress.

Flowing water, the principal agent for erosion in the Grand Canyon, causes it to deepen. Since the Colorado River encountered and became entrapped by hard rocks, it has eroded not laterally but downward. The Grand Canyon is being widened by several processes: mass

Figure 1 (left): Aerial view looking upriver near the Colorado River, Mile 214 Western Grand Canyon. Note large broken ridge in the contor of the canyon. Canyon walls are 1,500 meters high.

Figure 2 (right): Aerial photo looking upriver near Colorado River Mile 217. Landslide involves rocks that have slid down over 450 meters for nearly three kilometers. Note layers of rock in slide are rotated backward against parent cliff.

wasting, which is the movement of a large amount of rock during landslides or earthquakes; rockfalls; and sheetwash debris flows, in which major amounts of debris move rapidly in the wake of a downpour. Canyon widening is also noticeable where tributary streams join together, because in that junction, the valley wall is being attacked by running water from two different directions. Local erosion of softer rock often undercuts harder, cliff-forming rocks, which increases the likelihood of landslides or rockfalls.

The Role of Rockfalls and Debris Flows

One of the erosional processes largely responsible for the widening of the Grand Canyon is rockfalls. Rockfalls and debris falls include material that falls from vertical or overhanging cliffs, usually during storms. Rocks that fall or tumble down canyon slopes or tributary riverbeds are crushed and broken into smaller rocks and sand. When water flows down a tributary, it serves mainly as a conveyer belt to the Colorado River, carrying material delivered to it by sheetwash, rockfalls, and other forms of erosion. The Colorado River is the master conveyer belt. Before dams were constructed, the river continuously carried tons of sand and silt, minute by minute, day by day, to the Gulf of California.

What Happens to Debris of Different Sizes?

Most boulders never reach the river, and the few that do have usually traveled less than a mile. Boulders that have traveled longer distances have usually been transported by a flash flood or debris flow. Debris flows are common in the Grand Canyon and consist of a slurry of mud, sand, gravel, and boulders mixed with a little water. Gravity is the force that makes the mass move, and water is the lubricant. Debris flows travel at high speeds, often faster than a person can run. The density of these flows allows boulders as large as cars to roll along with ease. Debris flows dump large quantities of material into the Colorado River, where the larger boulders come to rest. Only the finer material is carried on downriver, because the Colorado River is not a fast-flowing stream and has a low gradient, so it's much less powerful than a flash flood or a debris flow rushing down a tributary. The river must flow over the boulders that it cannot move, which creates a rapid. Indeed, most rapids of the Colorado River in Grand Canyon are located where tributary streams enter the river and deposit boulders. Some rapids are formed by bedrock or large rockfalls, but these are not common. Eventually, over hundreds to thousands of years, the boulders are worn down to a size small enough to be transported farther downstream.

The Role of Landslides

Landslides are another major weathering process that widens canyons. In some areas of the Grand Canyon, especially in the western third, landslides have occurred on a large scale. Some are as much as one mile long and 900 meters feet high. Here, whole sections of rock have broken away from cliffs along natural joints or cracks and have slid downward and rotated backward towards the parent wall from which they came, the way a person slumping against a wall or slipping on ice ends up on his or her back. It is not certain if these landslide masses moved gradually downward or collapsed all at once in a major catastrophic event. But earthquakes have produced several major faults in the western Grand Canyon, and some of the bigger earthquakes probably initiated the landslide events.

Figure 3: Aerial view looking northeast up 205 Mile canyon, Western Grand Canyon. This is a fault-controlled canyon several kilometers long. Note straight fault line in canyon.

As a landslide mass slides down, the lower crumpling section tends to slow the slide to a gradual downward creep over a long but intermittent period of time, typically thousands of years. A landslide mass may gradually creep downward during long wet periods or stormy conditions. As a result of landslides, the western Grand Canyon is wider than the eastern Grand Canyon, where the faults are less prominent and large earthquakes less common.

Canyons Widen When Tributary Streams Grow Longer

Headward erosion, in which the higher, originating end of a river wears away the rock around it, lengthens tributary streams. This process also widens canyons. Some of the tributaries within the Grand Canyon have eroded headward along faults from the Colorado River into the plateau rims. The faults have broken and weakened the rock, allowing running water to establish a drainage that erodes quickly headward. Tributary canyons of this type are called fault-controlled canyons and account for about thirty-five percent or more of the tributary canyons in the Grand Canyon.

The Role of Springs

To a lesser extent, the action of springs also lengthens a tributary canyon, especially where they emerge at the head of a tributary stream. Water from a spring dissolves the cement that holds the rock together, causing the rock to weaken and crumble. These deteriorated rocks easily erode and undermine the rocks above, creating an overhang or amphitheater where landslides, rockfalls, and waterfalls are likely to occur, causing headward erosion.

The Role of Climate

As an agent of erosion, climate also sculpts the canyon. In humid or tropical regions, deep chemical decay of rocks creates soils and is considered one of the outstanding features of weathering. Chemical weathering is any process that dissolves a rock. In the Grand Canyon, the climate is arid to semi-arid and chemical weathering does not generate much soil, but it does help weaken, fragment, and decompose bedrock at and near the canyon rim, and aids in the general erosion of bedrock into smaller rocks and soil. Meager soil does develop on the surrounding higher plateaus, where precipitation is more common.

A striking difference between humid and arid climates can be seen in the character of valley or canyon walls. In humid areas, valley slopes or walls are clean of fragmented rock material called talus. This is because of deep chemical decay, the restriction of erosion by vegetation, and the lack of freezing and thawing. In the Grand Canyon, especially at higher elevations, valley slopes are typically littered with boulders and rocks of all sizes, due mostly to the action of freezing and thawing. In the winter, when snow melts, water in cracks in the rock freezes at night, wedging rocks apart. Eventually the rocks fall.

Figure 4: View looking down South Separation Canyon to the Colorado River (not visible) and northeast up North Separation Canyon, Western Grand Canyon. The separation fault controls the headward erosion of both canyons for several miles as fault-controlled canyons.

The influence of climate on different rock types also depends on their specific characteristics. For example, in humid areas limestone is usually considered a "weak rock," but in arid areas such as the Grand Canyon, it is considered a "strong rock" and commonly forms a cliff. The difference lies in the fact that limestone is highly susceptible to being infiltrated by water. Water contains carbonic acid, which pits and corrodes the surface of limestone, weakening it. People who compare the topography of arid and humid regions are usually impressed by the greater angularity of slopes in arid regions.

What Would an Eastern Grand Canyon Look Like?

What would the Grand Canyon look like if it were located in the humid eastern United States? The differences between the south- and north-facing sides of the canyon offer clues. The south-facing slopes and cliffs on the north side of Grand Canyon are generally less steep than the opposing north-facing slopes of the south rim. There are a number of explanations. North-facing slopes have a longer snow cover, experience fewer days of freeze and thaw, retain soil moisture longer, and generally have a better vegetative cover, all of which are likely to result in less active erosion. Since these conditions are more typical of the humid East Coast, such a canyon would be likely to have walls that slope more gently and have more plant growth.

The Ongoing Debate

Geologists know that the formation of the Grand Canyon is much more complicated than was once thought, and they are gradually finding more evidence to support new theories and ideas on how various parts of the Colorado River and the canyon came into being. Having mapped the Grand Canyon, I have come to believe that the canyon itself may not hold the answer. Many of the rocks that may have held clues as to its formation have largely been removed by erosion during the last few million years. The vast region of the Grand Canyon and surrounding plateaus has not been fully explored by geologists, and there are only a few geologists working in the area today. For these reasons, much is yet to be discovered. ☻

Death Among the Dunes: A Dinosaur Murder Mystery

Lowell Dingus and David B. Loope

Despite all the new high-tech tools available to modern detectives, some murder cases are still difficult to solve. Imagine trying to crack a case more than 70 million years old involving hundreds of mysterious deaths in Central Asia. It's not easy, but it is an important part of a joint American Museum of Natural History/Mongolian Academy of Sciences paleontological expedition.

Lowell Dingus is a Research Associate in the Department of Vertebrate Paleontology at the American Museum of Natural History. David B. Loope is a Professor in the Department of Geosciences at the University of Nebraska, at Lincoln.

Ankylosaur fossil uncovered at Ukhaa Tolgod in Mongolia.

Our expedition's fossil discoveries have been widely chronicled. Hundreds of exquisitely preserved mammals, lizards, birds, and other dinosaurs have been retrieved from the harsh and desolate expanses of rusty red sandstone scattered across Mongolia's Gobi Desert. Many specimens represent new species previously unknown to science, and some fill important gaps in our knowledge about the evolution of birds and mammals. In the wake of discovering such a wealth of paleontological treasures, some questions naturally arise. What was the world like in which these ancient animals lived? How did they die, and how were their fragile skeletons so perfectly preserved in this remote region?

The clues for solving these mysteries lie in both the fossil skeletons and the rocks that contain them. Most fossils collected by our expedition come from a locality that our crew discovered in 1993 called Ukhaa Tolgod. This site has already yielded the richest assemblage of Mesozoic fossil vertebrates known anywhere in the world. Skeletons range in size from twenty-foot-long ankylosaur specimens complete with almost every piece of bony armor to two-inch-long mammal skeletons preserving microscopic skeletal structures such as ear bones. Some clearly died in natural lifelike positions, such as the dinosaur Oviraptor brooding her nest of eggs. This unusual degree of preservation indicates that many of these animals were killed and buried quickly, possibly by some kind of catastrophic event. Otherwise the bones would have been scavenged by other animals or scattered by the elements. But what events were responsible for such mass mortality and swift burial? Ultimately, the rocks that entombed the fossils hold the keys to unlocking this ancient riddle, because they contain the clues for reconstructing the area's ancient environment.

Lowell Dingus records findings in his field notebook at Ukhaa Tolgod, Mongolia.

Today, the Gobi encompasses windswept basins and rugged ranges. Impressive fields of sand dunes migrate across the bottom of the valleys. Violent sandstorms and thunderstorms occasionally interrupt our treks across the sparsely-vegetated slopes and ravines in search of fossils. A tent can be sent suddenly careening across the landscape, and the only protection from being sandblasted by windblown grit is to run for the cover of the expedition's trucks. Amazingly, the rocks at Ukhaa Tolgod tell us that not a lot has changed in the last 70 or 80 million years.

We know this because some of the reddish brown ridges at Ukhaa Tolgod are composed of sandstone layers that contain thin, tilted laminations (compressed layers of rock). These

laminations tilt consistently at about 25 degrees to the northeast. Similar laminations are found in the modern dunes of the Gobi and other deserts. They form when sand grains, blown by strong winds, bounce up the gently sloping windward side of the dune and tumble down the steeply sloping leeward side. The small ripples that contain the laminations are preserved when more sand tumbles down the leeward side as the dune migrates in the direction that the wind blows. These tilted laminations of sand are called crossbeds because they cut across thicker beds of sandstone. The consistent northeast tilting of the laminations represents evidence that the wind blew and the dunes migrated from the southwest to the northeast at Ukhaa Tolgod during the late Cretaceous Period.

Consequently, one hypothesis put forward to explain the sudden death and rapid burial of animals at Ukhaa Tolgod suggested that they died and were rapidly entombed during massive sandstorms. At first glance, this appears to

make a lot of intuitive sense. However, two major problems exist. First, unless the animals were already sick or injured, why would they let themselves be buried by sand to the point of suffocation? We have searched the travel literature and can find no account of animals being buried alive by sandstorms in the deserts of Central Asia or Arabia. So, this does not appear to happen frequently in the modern world. Second, hardly any of the fossil skeletons found at Ukhaa Tolgod were collected from the distinctly crossbedded sandstones generated by dunes. Occasionally, such crossbedded laminations were deformed into depressions, which look like a stack of bowls in cross-section and appear to represent footprints of the dinosaurs and other animals that walked across the dunefield. But these are footprints, not fossil skeletons.

Essentially all of our fossil skeletons were found in sandstone layers free of distinct internal structures like the tilted laminations typically found in dunes. Instead, they are contained in

David Loope digging into a cliffside at Ukhaa Tolgod.

massive, even-textured layers of sand. We began to wonder how these layers formed. When we looked closely, we noticed that these massive sandstones occasionally contained large isolated pebbles and cobbles. Clearly these were too heavy to be blown into the deposit by wind, so wind had not generated the fossiliferous layers of sand. But what had? Did they represent the filling of stream channels that had once run through the ancient dunefield? At first we weren't sure.

To complement our studies of the ancient rocks and fossils from the Gobi, we also work in the Nebraska Sand Hills—a large, relatively young dunefield in the western part of the state, presently stabilized by prairie vegetation. Sinuous dunes rise as high as 120 meters above the interdune valleys, and their leeward slopes dip at angles of 25–30 degrees, just like the crossbedded sands at Ukhaa Tolgod. Most rain falling in the Sand Hills infiltrates the permeable sediments and is slowly released from the groundwater reservoir by springs. However, heavy rainfall during summer thunderstorms can trigger sudden avalanches of wet sand, which are called debris flows. One rancher showed us a striking photo of his pickup half buried by sand during a July 1991 cloudburst. Other residents tell of their calving sheds—built on the leeward side of big dunes for protection from spring winds—being rapidly filled by sand slurries. In other areas, such flows are known to have carried large pebbles and cobbles that "float" in the sand as it moves, just like those found at Ukhaa Tolgod.

Are the Nebraska debris flows closely analogous to the 80-million-year-old massive sandstones that contain the fossils at Ukhaa Tolgod? Did the wet sand act like wet cement to trap the ancient inhabitants of Ukhaa Tolgod? We have not yet found any buried skeletons in Nebraska representing the recent prairie fauna, such as bison, coyotes, or prairie dogs. One possible reason is that we have not dug enough trenches. Even at Ukhaa Tolgod, where articulated vertebrate fossils are more concentrated than at any other known Mesozoic site, we never dig for fossils blindly. Instead, we look for bones exposed by erosion before starting to excavate. It also appears to us that, although the modern process may be similar to the ancient, the Nebraska debris flows are too small to overwhelm large animals. Why else have we not met a rancher who has lost a cow to a debris flow?

Individual flows in the Sand Hills deliver a lobe of sand as large as 300 meters long, a meter and a half thick, and fifteen meters wide. Their size appears to be limited by meteorological factors. The Nebraska dunes lie in an area of low relief northwest of the path taken by tropical storms. No storm delivering more than eighteen centimeters of rain in a twenty-four-hour period has been recorded in the Sand Hills since 1950. One hundred miles north of the Sand Hills, in Rapid City, South Dakota, however, the configuration of the Black Hills mountain range allowed twenty-five centimeters of rain to fall in six hours in 1972. We know that mountains rimmed Mongolia's basins in the Late Cretaceous and that these basins lay just west of an unusually warm Pacific Ocean. It is quite possible that the ancient Gobi dunes were occasionally drenched by rainstorms that would dwarf those that hit western Nebraska.

In 1996, we found evidence that the ancient dunes at Ukhaa Tolgod migrated only intermittently. It was not until 1998, however, that we realized that this might help us interpret how the fossils were preserved. Our observations at Ukhaa Tolgod show that during periods of dune stability, sand about one meter beneath the surface of the dune slopes became cemented by calcium carbonate, forming thin, hard layers of what is called caliche. These cemented layers greatly slowed the infiltration of rainwater during heavy storms and may have played an important role in triggering the debris flows. Many of the recent devastating debris flows on the mountain slopes of coastal California and Central America were triggered by the development of a perched water table during heavy rains when infiltrating water moved through the soil but was impeded when it reached underlying bedrock. Then, the soil becomes fully saturated, and part of its weight is "lifted" by water between sediment grains. This decreases the friction between the soil and the bedrock that, under dry conditions, keeps the steep slope stable. In Mongolia, during heavy rains, the water-saturated sand above the cemented zone would have broken loose and moved down the long dune slope at high speed. No caliche layers are present in the Sand Hills, which could explain the absence of large debris flows.

The two kinds of sandstone preserved at Ukhaa Tolgod probably represent two distinct climatic regimes that prevailed at different times. Crossbedded dune sand, where we find rare sets of fossilized tracks, accumulated during arid intervals when plant and animal life was less abundant. Unstructured sandstones, containing articulated skeletons, represent debris flows that occurred when atmospheric circulation had changed sufficiently to bring more rainfall to the dunefield. The wetter conditions led to dune stabilization, caliche formation, and a sharp increase in the abundance and diversity of plant and animal life. The key elements for our story—steep dune faces, caliche, heavy rainfall events, and abundant animal life—were all required for the lethal fossilization to take place. After one or more deadly flows, the climate grew drier, and the fossil-bearing sediments were overrun as the sparsely vegetated dunes started to migrate again.

The 80-million-year-old murder mystery at Ukhaa Tolgod has yet to be completely solved. But we think we are on the right track. It's clear that the dinosaurs, mammals, and lizards that inhabited Ukhaa Tolgod did not live in an unbearably hostile environment. Although their lives were occasionally punctuated with violent tempests and perilous sand slides, they roamed across majestic fields of stately sand dunes that, although probably hot and arid at times, supported adequate vegetation and abundant prey. 🌑

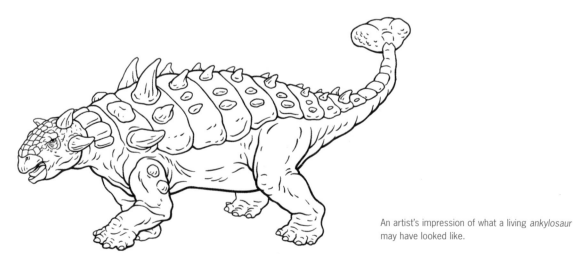

An artist's impression of what a living *ankylosaur* may have looked like.

A view of the glacier-clad summit of Mt. Rainier.

Mapping Mt. Rainier

Tom Sisson hadn't decided which of the sciences to pursue until he took a geology course taught by a professor who "besides being interested in the material, was in his mid-60s—infinitely old to me at the time," he recalls, chuckling. "We would go on field trips and this guy would go walking briskly up these hills to various outcrops, and leave all these college students gasping in his wake. And I thought, 'Well, this guy's doing something right.'" Already an experienced ice climber, Sisson went on to become an accomplished mountaineer and a geologist with the Volcano Hazards Program at the U.S. Geological Survey.

For six years, Sisson has been studying the geological history of Mt. Rainier, a volcano in the Cascade Range of Washington State. When the project is completed he will have pieced together a three-dimensional map of Mt. Rainier

and a detailed understanding of its complex and tumultuous past. "Studies had been made of mudflows after the end of the ice ages, in the last 10,000 years, and also of ash deposits blasted up again since then. But the volcano has been around an awful lot longer," explains the geologist. "No one had tried to look at Rainier's overall growth from its inception as far back as one could infer, which is about half a million years."

In order to do so, Sisson has surveyed every ridge and rock face of Mt. Rainier, a mountain on which it is notoriously difficult to get around. "The main hazard on Rainier is that it's tall, with a lot of glaciers. You have to avoid falling off a ridge or a crevasse, a deep crack in the rock," he explains matter-of-factly. "I use standard mountaineering and belaying techniques. You also have to pay attention to where you are at what time of day, because the steeper areas spit rocks when they get warm." Snow and ice wedge rocks apart as temperatures drop; as the rocks thaw, they loosen and fall, dislodging larger rocks as they tumble. "Usually they are fist- to head-size, but they can be as big as cars," says Sisson, "so you don't want to be there." Then there's the weather, about which the geologist suggests, "Take the long view. If a snowstorm is coming, go do something else."

An aptitude for mapping is another definite asset. "You always have to know pretty accurately where you are so that whatever you see you can plot on the map," Sisson explains. "You take people outside and wander around for a bit and ask them where they are, and some people can tell and some can't. I happen to be one of those who can." He is also highly observant, which he attributes to a combination of practice and aptitude. And he likes puzzles. "For example, you're walking up a ridge, and you walk over one kind of lava. A little further on, you walk over another kind of lava. Then another like the first stuff you walked over. What's going on? Is it a series of flows that are stacked up and alternating? Or did a younger lava flow drape over the older one and then erode through? You've got to wander around to figure out which answer is right. It's like a big jigsaw puzzle." Every detail is meticulously noted in a small, waterproof notebook, then translated, centimeter by centimeter, into a blueprint of the events which shaped the mountain. Sisson also collects rock samples to take back to the laboratory, where their ages are determined by isotope-dating techniques.

In addition to mapping lava flows, Sisson maps pyroclastic flows—high-speed avalanches of hot ash, rock fragments, and gas—dykes, faults, glacial moraines, pumice, and ash. In other words, "whatever's out there. When you're mapping a volcano, you're trying to figure out a number of things, the first of which is how it has behaved in the past, because that's the best guide to how it's going to behave in the future. What kind of lava has it produced? How much? Where did it go? How frequently does it produce flows? How hot were they? The same goes for pyroclastic flows or ash flows or lava domes."

This new interest in Rainier's past is attributed to the recent, rapid increase in population density around its flanks, and new evidence shows that Rainier has been much more active than scientists previously thought. More than once, enough molten rock has spilled from the volcano to bury an area the size of Tacoma and Seattle almost ten feet deep. Particularly alarming is the discovery that large mudflows have occurred every 500 to 1,000 years. Scientists fear that a lahar—a fast-flowing river of mud, rocks, blocks of ice, and trees—could surge off the mountain with little warning, if any. Over the past 5,000 years, dozens of such flows may have extended beyond the base of the mountain, and at least six have flowed down river valleys and reached the ocean.

Mt. Rainier is often called the most dangerous volcano in the United States, but not by Sisson. "I've never said that and I wouldn't, because I don't find it a very helpful thing to say. The reason people say so is not because it erupts so frequently but because so many people live around it." Another cause for concern is the fact that much of Mt. Rainier's rock is weak and crumbly. "Rainier, like most other volcanoes, supports what's called a hydrothermal system," Sisson explains. "The interior of the volcano is hot because magma has moved through it. Rain or snow falls and seeps into the interior. The complicating factor is that both magma and the rock it solidifies into contain sulfur. The sulfur gets leached out of rock or boils out of magma, is absorbed in the water, and makes the water acidic. That hot acidic water circulating through

the volcano etches and weakens rock, turning it into clay." However, Sisson notes, "the water doesn't circulate uniformly." In fact, the bulk of Mt. Rainier is unaffected by this hydrothermal activity, which is why the geologist objects to dramatic language comparing the mountain to a house being eaten by termites or stewing in its own juices. "Most of it is stable," says Sisson firmly, adding, "for a volcano. Volcanoes are inherently unstable." Mt. St. Helens, he points out, had little hydrothermal activity in the areas of rock that were blown out during the 1980 eruption.

Mt. Rainier's last eruption, a small one, happened in 1884. The last major eruption took place about 1,100 years ago, and the one before that approximately 2,300 years ago.

Until recently, the volcano was perceived as being old and near-dormant—but not any longer. What impresses Sisson the most about Mt. Rainier "is that it's just cranking along the way it has for half a million years. There's no indication that this volcano is dying out. So we can expect that this volcano is going to continue producing lava flows that go for a handful of miles, and pyroclastic flows that melt glaciers and produce volcanic mudflows, and there are going to be people living in those places. If we can warn them, then they'll have a chance to evacuate. Otherwise," he concludes, "many hundreds of people may lose their lives." ☮

Dacite columns that formed tens of thousands of years ago when a lava flow cooled rapidly against a glacier.

James Hutton: The Founder of Modern Geology: Profile

View of Siccar Point, Scotland.

James Hutton (1726–1797), a Scottish farmer and naturalist, is known as the founder of modern geology. He was a great observer of the world around him. More importantly, he made carefully reasoned geological arguments. Hutton came to believe that the Earth was perpetually being formed; for example, molten material is forced up into mountains, eroded, and then eroded sediments are washed away. He recognized that the history of the Earth could be determined by understanding how processes such as erosion and sedimentation work in the present day. His ideas and approach to studying the Earth established geology as a proper science.

In the late eighteenth century, when Hutton was carefully examining the rocks, it was generally believed that Earth had come into creation only around six thousand years earlier (on October 22, 4004 B.C., to be precise, according to the seventeenth century scholarly analysis of the Bible by Archbishop James Ussher of Ireland), and that fossils were the remains of animals that had perished during the Biblical flood. As for the structure of the Earth, "natural philosophers" agreed that much bedrock consisted of long, parallel layers which occurred at various angles, and that sediments deposited by water were compressed to form stone. Hutton perceived that this sedimentation takes place so slowly that even the oldest rocks are made up of, in his words, "materials furnished from the ruins of former continents." The reverse process occurs when rock exposed to the atmosphere erodes and decays. He called this coupling of destruction and renewal the "great geological cycle," and realized that it had been completed innumerable times.

Hutton came to his chosen field by quite a roundabout route. Born in Edinburgh in 1726, he studied medicine and chemistry at the Universities of Edinburgh, Paris, and Leiden, in the Netherlands, and then spent fourteen years running two small family farms. It was farming that gave rise to Hutton's obsession with how the land could hold its own against the destructive forces of wind and weather he saw at work around him. Hutton began to devote his scientific knowledge, his philosophical turn of mind, and his extraordinary powers of observation to a subject that had only recently acquired a name: geology.

Around 1768 he moved to Edinburgh, where a visitor a few years later described his study as "so full of fossils and chemical apparatus that there is hardly room to sit down." In a paper presented in 1788 before the Royal Society of Edinburgh, a newly-founded scientific organization, Hutton described a universe very different from the Biblical cosmos: one formed by a continuous cycle in which rocks and soil are washed into the sea, compacted into bedrock, forced up to the surface by volcanic processes, and eventually worn away into sediment once again. "The result, therefore, of this physical enquiry," Hutton concluded, "is that we find no vestige of a beginning, no prospect of an end." Relying on the same methods as do modern field geologists, Hutton cited as evidence a cliff at nearby Siccar Point, where the juxtaposition of vertical layers of gray shale and overlying horizontal layers of red sandstone could only be explained by the action of stupendous forces over vast periods of time. There Hutton realized that the sediments now represented by the gray shale had, after deposition, been uplifted, tilted, eroded away, and then covered by an ocean, from which the red sandstone was then deposited. The boundary between the two rock types at Siccar Point is now called the Hutton Unconformity.

The fundamental force, theorized Hutton, was subterranean heat, as evidenced by the existence of hot springs and volcanoes. From his detailed observations of rock formations in Scotland and elsewhere in the British Isles, Hutton shrewdly inferred that high pressures and temperatures deep within the Earth would cause the chemical reactions that created formations of basalt, granite, and mineral veins. He also proposed that internal heat causes the crust to warm and expand, resulting in the upheavals that form mountains. The same process causes rock stratifications to tilt, fold and deform, as exemplified by the Siccar Point rocks.

Another of Hutton's key concepts was the Theory of Uniformitarianism. This was the belief that geological forces at work in the present day—barely noticeable to the human eye, yet immense in their impact—are the same as those that operated in the past. This means that the rates at which processes such as erosion or sedimentation occur today are similar to past rates, making it possible to estimate the times it took to deposit a sandstone, for example, of a given thickness. It became evident from such analysis that enormous lengths of time were required to account for the thicknesses of exposed rock layers. Uniformitarianism is one of the fundamental principles of earth science. Hutton's theories amounted to a frontal attack on a popular contemporary school of thought called catastrophism: the belief that only natural catastrophes, such as the Great Flood, could account for the form and nature of a 6,000-year-old Earth. The great age of Earth was the first revolutionary concept to emerge from the new science of geology.

The effect that this portrait of an ancient, dynamic planet had on the thinkers who followed in the next century was profound. Charles Darwin, for example, was well acquainted with Hutton's ideas, which provided a framework for the eons required by the biological evolution he observed in the fossil record. English geologist Sir Charles Lyell, who was born the year Hutton died and whose influential book Principles of Geology won wide acceptance for the Theory of Uniformitarianism, wrote, "The imagination was first fatigued and overpowered by endeavouring to conceive the immensity of time required for the annihilation of whole continents by so insensible a process." The "ideas of sublimity" awakened by this "plan of such infinite extent," as Lyell referred to it, inspired not only Hutton's contemporaries, but generations of geologists to come. 🌓

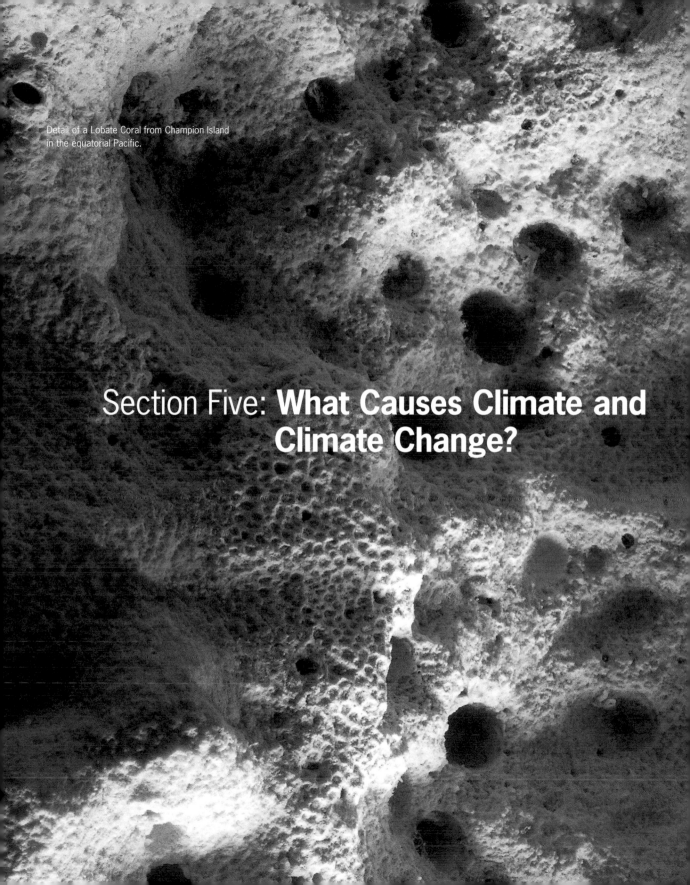

Detail of a Lobate Coral from Champion Island in the equatorial Pacific.

Section Five: **What Causes Climate and Climate Change?**

Introduction Edmond A. Mathez

Climate is the long-term state of weather. Understanding how the climate system works has become important because human activity can modify it, possibly in profound ways. But climate is not so easy to understand. One reason is that it is largely controlled by the interaction of the atmosphere and ocean and the ways they circulate, which are themselves complex. The Earth receives most of the Sun's heat in the equatorial regions. Atmosphere and ocean circulation serve to transfer this heat to the polar regions, and this is what fundamentally drives the climate system.

Numerous factors influence climate. Some operate over millions of years; others operate over shorter periods, such as a year or less; and some of these factors affect each other. Because of the important coupling of the ocean and climate, the first essay in the section, by Martin H. Visbeck, explains how the ocean circulates, regulates climate, and causes it to change over time spans ranging from seasonal to 10,000 years and longer.

One of the clearest examples of the interaction between the atmosphere and ocean is the phenomenon of El Niño, so this is a fitting subject for the second essay by Mark A. Cane. El Niño is the second most important periodic change in climate, after the seasons. It refers to a warming of the normally cold waters of the equatorial Pacific Ocean off the coasts of Peru and Equador. In an El Niño year, the ocean currents and wind of the equatorial Pacific, which normally move westward, reverse direction and flow to the east, transporting warm water of the western Pacific to the east. The shift in wind and ocean current direction is accompanied by changes in the atmospheric pressure between South America and the India-Australia region, known as the southern oscillation. The whole phenomenon is called

ENSO, for "El Niño Southern Oscillation." ENSO affects climate around the globe. We need to pay attention to it because it influences everything from local weather to food production. One of the most amazing examples of this has come from Dr. Cane's research. He discovered that the harvest of maize (corn) in Zimbabwe follows closely, by several months, the temperature of the ocean water off Ecuador—when the water is warm, Zimbabwe experiences dryer conditions and lower maize yields. Zimbabwe is in southern Africa, nearly a third of the way around the globe from the eastern Pacific.

The amount of greenhouse gases in the atmosphere influences the climate. The most important greenhouse gases are water vapor and carbon dioxide (CO_2). These gases trap radiation that emanates from the Earth's surface, acting as an insulating blanket. If all the carbon on Earth were in the form of carbon dioxide in the atmosphere instead of in the form of rock, Earth might be much like Venus—a lifeless planet whose surface temperature is about 460°C. We shall explore this further in Rachel Oxburgh's essay in Section Six. On the other hand, although water vapor is also a greenhouse gas, clouds actually reflect more sunlight back into space than, say, land or water, and thus act to cool the Earth. This is also true of ice. Dust and manmade pollutants in the atmosphere also tend to cool the Earth, for the same reason.

The debate about global warming concerns the effect on climate of the increasing amounts of carbon dioxide being pumped into the atmosphere by the burning of fossil fuels. This and the change in climate over the last century are the subjects of Charles F. Keller's essay. No one disputes that the Earth has been warming and that atmospheric CO_2 content has been rising. However, one of the things we've learned is that there is no simple relationship between temperature and atmospheric CO_2 content because there are so many other factors at work. We really don't understand how the build-up of CO_2 will influence our climate. To paraphrase Wallace Broecker, one of the eminent scientists working on the problem, it's like poking an unknown and potentially dangerous animal with a big stick.

Among the other long-term factors that influence climate are the orbital characteristics of the Earth, which cause a change in the amount of solar radiation reaching our planet. Firstly, the Earth's orbit around the Sun is elliptical, and the shape of the ellipse varies. Secondly, the orbital plane—the tilt of the Earth's axis relative to the Sun—oscillates back and forth. In addition, the Earth's axis of rotation itself rotates around another axis—in other words, it "wobbles" like spinning tops usually do. The possibility that these orbital parameters could combine to cause the ice ages of the last two million years was first explored by the Serbian mathematician, Milutin Milankovitch, whose profile is part of this section. The way that each of these parameters influences climate depends on where the major land-masses are and how the oceans are circulating. Nonetheless, the orbital parameters are clearly related to the ice ages and interglacial periods of the last two million years, and Milankovitch cycles, as they have become known, have been recognized in the geologic record of the more distant past.

How have we come to our present under-standing of the global climate system? It has certainly not been only with observations of the current state of our planet. Rather, much of our insight has been gained by studying the past climate, or the paleoclimate. Paleoclimate is derived from records that include everything from ocean and lake sediment cores, to corals, tree rings, ice cores and peat bogs. The thickness of annual growth rings of trees is sensitive to moisture and other factors. Because they are stationary, trees obviously record only local conditions. But they can be dated very precisely, and the examination of many different trees can provide information on conditions over wide regions. This is Gordon Jacoby's research as described in this section's case study.

Ice cores from the polar regions have given us great insight into how the climate system works. The final essay of this section, by Paul A. Mayewski, describes how ice is used to determine paleoclimate. One of the most exciting records comes from ice cores drilled through the entire Greenland ice cap, some 3,000 meters thick. These cores have recorded all sorts of things: volcanic eruptions, the beginning of the Industrial Revolution, a 1908 meteorite impact, even the passage of the 1974 Clean Air Act. But they also provide a detailed record of climate back to about 110,000 years. The record contains some unexpected and profound events. One of them is the Younger Dryas, a cold period that began about 13,000 years ago, after the last glaciation, and lasted for 1,300 years, killing the forest that covered much of North America and northern Europe. The onset of this dramatic and enormous shift in climate occurred over just a few decades, and, similarly, terminated over just a few decades. This is unlike anything humanity has ever experienced. The rapid change implicates the ocean as the cause of the Younger Dryas, because only a dramatic change in ocean circulation could move so much heat around in such a short period of time to force such rapid climate change. The Earth's climate is dynamic, chaotic, and complex, and we have much to learn about it. Because our activities may dramatically influence climate, and because climate in turn may have a significant effect on how we live, climate change policy will continue to be part of the political debate. ❻

To explore what causes climate and climate change, I pose the following questions:

What role do oceans play in climate change?

Martin H. Visbeck, an Associate Professor in the Department of Earth and Environmental Science at Columbia University's Lamont-Doherty Earth Observatory, explores how the world's oceans regulate climate.

What is El Niño and how can we predict it?

Mark A. Cane, the G. Unger Vetlesen Professor of Earth and Climate Sciences at Columbia University's Lamont-Doherty Earth Observatory, describes the skepticism of the scientific community when his computer model first successfully forecasted an El Niño event.

Why is global warming such a hot topic?

Charles F. Keller, the Center Director of the Los Alamos National Laboratory branch of the University of California's Institute of Geophysics and Planetary Physics, explains natural and human factors which warm the planet and discusses modeling techniques for predicting future consequences of global warming.

How can ice record climate change?

Paul A. Mayewski, the Director of the Climate Studies Center and a Professor in the Institute for Quarternary Studies and Department of Geological Sciences at the University of Maine, explains how ice cores drilled in Antarctica and Greenland record events that affect the global climate.

138 139

Scientists carefully extrude an ice core from its barrel.

The Oceans' Role in Climate

Martin H. Visbeck

A Numerical Portrait of the Oceans

The oceans of the world cover nearly seventy percent of its surface. The largest is the Pacific, which contains fifty percent of the volume of all the oceans combined, followed by the Atlantic and Southern Oceans. The total mass of the ocean is small compared to that of the solid Earth (10,000 times less) but large when compared to that of the atmosphere (300 times larger). A typical ocean depth is 3,000 meters, which is about 0.05 percent of the Earth's radius, so the ocean is a shallow

Martin H. Visbeck is an Associate Professor in the Department of Earth and Environmental Science at Columbia University's Lamont-Doherty Earth Observatory.

layer of fluid covering much of the Earth's crust.
The oceans account for ninety-seven percent
percent of the total water on Earth. Seawater
itself is a mixture of ninety-six percent fresh
water and 3.5 percent dissolved salts. Its
temperatures range from the freezing point of
ocean salt water (-1.8°C) in the polar oceans to
a maximum of 34°C in the tropics, with an
overall average temperature of 3.8°C. About five
percent of the ocean's surface is covered by sea
ice. The viscosity of the oceans, or their
resistance to flow, is much lower than that of the
solid Earth, but significantly higher than that of
the atmosphere. This means that ocean currents
travel several orders of magnitude faster than
the solid Earth, but considerably slower than the
atmosphere. Ocean currents travel at speeds of
up to three meters per second, while winds
move at least ten times faster and the solid
Earth about a trillion times slower.

Ocean Dynamics

Ocean currents can be driven by the wind
The wind blowing over the surface of the water
drives the ocean's major surface currents.

These winds in turn are driven by atmospheric
circulation, generated by unequal temperatures
in the atmosphere. The main features of this
wind-driven surface circulation are large, roughly
circular current systems, called gyres. Gyres
are found in most major ocean basins (Figure
1). Driven by the prevailing wind systems and
deflected by continental boundaries and the
Coriolis force resulting from the Earth's rotation,
gyres help redistribute heat from the low
latitudes to the polar regions. Along the western
margins of the ocean basins, warm ocean
currents like the Gulf Stream transport heat
towards the poles. Along the eastern margins,
currents such as the California Current
transport cold water to the lower latitudes.

In the landlocked higher latitudes of the
Northern Hemisphere, the prevailing winds drive
smaller gyres that effectively redistribute heat
to the polar regions. In the Southern

Figure 1: Ocean Surface Currents. Surface currents are
driven by the winds, and in turn influence atmospheric
circulation. Surface currents form huge gyres–black arrows
are warm currents and blue arrows are cold currents.

Hemisphere, where no continents block ocean movement, strong westerly winds drive the largest flows of ocean water in the world around Antarctica. These flows move 180 million cubic meters of water per second. (In comparison, the Amazon River moves one million cubic meters per second.)

Variations in water density also drive ocean currents

The ocean is a layered system. Warm water lies close to its surface with cold water below. Winds, waves, and currents stir the ocean surface to form a mixed-layer a few tens of meters deep. In the low- and mid-latitudes, other layers lie below this warm mixed-layer. These include: intermediate water immediately below the mixed-layer, deep water, which extends from below the intermediate water to near the ocean bottom, and bottom water, which is in contact with the seafloor.

Surface seawater becomes denser as it cools or becomes saltier due to evaporation. It becomes lighter as it warms by heating due to the Sun or its salinity decreases due to the addition of fresh water from river outflows and rain. These changes in the density of seawater, even when small, drive the layered circulation of the ocean. There are only a few places on Earth where surface seawater becomes dense enough to sink to great ocean depths. This sinking water draws up surface waters from lower latitudes to replace it, and thereby generates movement on a large scale. In the Northern Hemisphere, the locations where surface water sinks are at the centers of the Greenland and Labrador Seas. In the Southern Hemisphere, surface water sinks to deep ocean depths along the shallow ocean margins around Antarctica, where dense water is produced by an intricate process involving heat loss to the atmosphere and interactions with sea ice. Since the density variations that drive the deep

circulation are due to differences in temperature (thermal) and salinity (haline), the density-driven ocean circulation is also referred to as thermohaline circulation. The resulting large-scale ocean circulation plays a fundamental role in the Earth's overall climate. Changes in ocean circulation can result in dramatic regional and global climate change.

How the Oceans Help Regulate Earth's Climate

The oceans modulate the planet's heat distribution

The Sun is the ultimate source of the energy that brings about atmospheric and oceanic circulation. Because of astronomical and atmospheric factors, tropical regions receive more of the Sun's energy over the course of a year than do regions at higher latitudes. However, latitude has much less of an effect on outgoing radiation, which remains relatively constant. This creates an imbalance at different latitudes between gains and losses of energy from the Sun. This radiative imbalance is balanced by heat flows from the equator towards the poles with the atmosphere and oceans. This heat transport is accomplished in about equal parts by the circulation of the atmosphere and that of the ocean. However, the transport within the ocean basins is not evenly split. Most heat is transported by the Atlantic Ocean.

The oceans modulate Earth's distribution of fresh water

The strong heating of the subtropical and tropical oceans causes evaporation at the surface. This moist air moves towards the equator, where it rises to the top of the lower atmosphere (to a height of about ten kilometers) and most of its water rains out. Strong rainfall reduces the salinity of the surface water. The generally poleward ocean

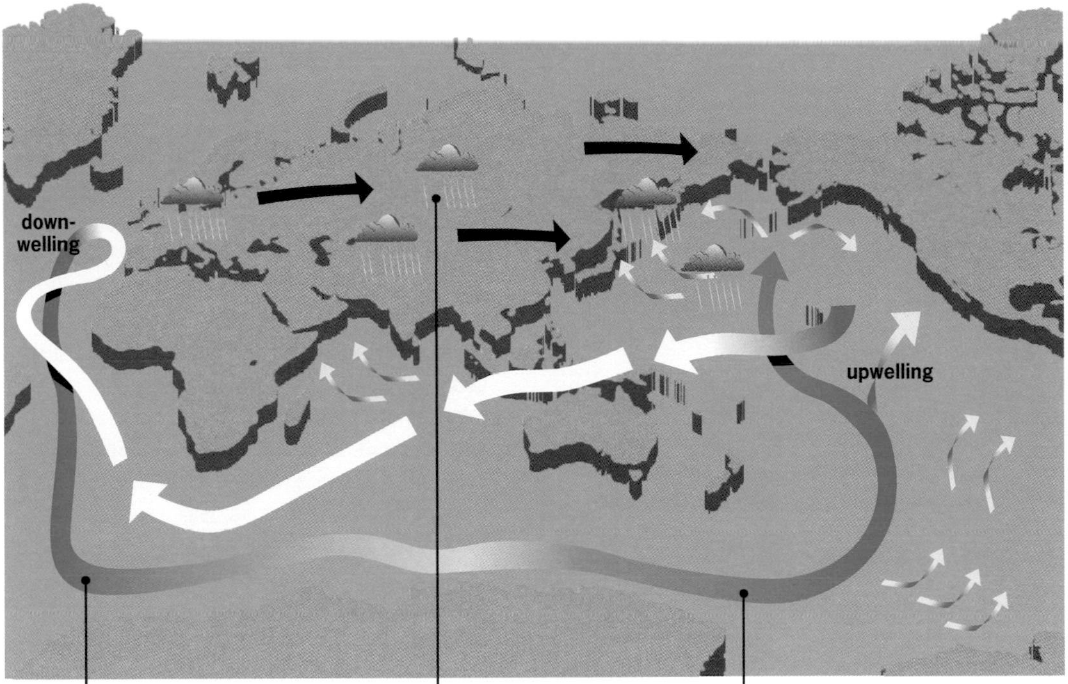

down-
welling

upwelling

Deep currents carry cold,
salty water from the
Atlantic to the Pacific.

Fresh water is
carried eastward
by rain clouds.

Water takes thousands of
years to circulate through
the oceans.

circulation transports the fresh water back to
the subtropical latitudes where it can evaporate
again. A similar hydrological cycle exists
between the subtropical and middle latitudes.
More precipitation occurs over the middle
latitudes and fresh water is returned to the
subtropics by oceanic flow.

How the oceans affect climate variability
Changes in the climate have different durations.
Strong climate variations occur every season.
Subtler but still noticeable changes occur on an
interannual scale; some last from a few
decades to a century or more. Dramatic
changes are linked to ice ages, which occur
over time scales greater than 10,000
years. The oceans affect all of these patterns
of variability.

Figure 2: Water warmed at the equator by the Sun flows into
the North Atlantic, where it is cooled and becomes more
salty because of evaporation. This cold, salty water sinks to
the seafloor and forms a huge undersea river. The deep
water flows through the oceans, welling up where the winds
push away warm surface water. This transfer of salty water is
balanced by fresh water evaporated from the Atlantic and
carried to the Pacific by the atmosphere. There, it falls as
rain, diluting the upwelling salty water with fresh water.

The seasonal cycle
The effect of the oceans in moderating
seasonal climate is the most familiar to us. The
enormous heat capacity of the oceans and of
large lakes, relative to that of the land, buffers
seasonal changes in solar radiation. During
winter, the ocean warms the lower atmosphere
above it, which results in milder temperatures
along the coasts and downwind of the major

oceans (along the west coasts of the United States and Europe, for example). In the summer, however, strong radiation heats the ocean more slowly than the land surface, so the atmosphere over the ocean remains cool. Climatologists note this difference by referring to "maritime" and "continental" climates.

These seasonal differences between air temperatures over land and water drive changes in large-scale atmospheric circulation. The most dramatic effects occur between the Indian Ocean and the Asian continent when winds called monsoons reverse seasonally. In the summer, air rises over the warm continent, which causes cooler, moist air masses to flow in from the Indian Ocean. As this air approaches the Himalayan mountains it rises, and it begins to rain over India. A similar phenomenon affects the southern United States when moist summer winds from the Pacific Ocean and Gulf of Mexico bring rain to the central U.S.

Variability on a scale of one to ten years

Not every winter is like the previous one; some are mild and some are harsh. Scientists refer to this as interannual climate variability. In the last century, climatologists have discovered several interannual climate phenomena, most of which are coupled to changes in the way the oceans store or transport heat.

The prime example of this interannual climate variability is called ENSO, which stands for "El Niño Southern Oscillation." In the tropical Pacific Ocean, warm temperatures near Indonesia cause the air to rise to great height. This is called atmospheric deep convection. At the same time, ocean currents bring cooler waters toward the equator along the coast of South and North America, thereby lowering the surface temperature. Consequently the atmospheric temperatures drop, and a large atmospheric circulation cell develops. Warm air

rises in the western tropical Pacific, near Indonesia, and sinks in the eastern Pacific, off the coast of Central America. Associated surface winds blow water from east to west, generating a large pool of warm water in the western tropical Pacific, between Australia and Japan. Sometimes, for reasons that are not yet entirely clear, the winds relax. If this happens, some of the warm water flows back eastward, carrying heavy rainfall with it. This reduces the temperature difference between the western and the eastern Pacific, so the wind relaxes even more. The warm water then flows all the way to the eastern Pacific, which suppresses the cold ocean currents along the coasts of the Americas. This phenomenon occurs to some extent every year around Christmas time and was named El Niño (after the Christ Child) by fishermen in South America who were not able to catch fish at that time, which normally thrive in the cold currents. However, every three to seven years, the changes in the tropical temperatures are large enough to cause shifts in the atmospheric circulation and rainfall patterns over much of the globe.

The term El Niño has come to be reserved for these large-scale events. Once the warm water has cooled, the westward wind starts to develop again, and the climate system returns to its normal state. Sometimes an El Niño event is followed by the movement of colder-than-normal surface waters across the Pacific towards Indonesia. This cold phase is called La Niña.

Changes in the tropical sea surface temperatures also change the north-south, or meridional, direction of atmospheric circulation. During El Niño, an area of the tropical Pacific much larger than normal is covered with warm water. This enhances atmospheric winds and the transport of heat towards the poles. The ocean outside of the tropics responds to those changes in the wind, and the ocean transport of heat also increases. However, the slow response time of the large oceans (remember, ocean currents travel at least ten times slower than atmosphere currents) can preserve the signal—the memory of unusual temperature patterns—for several years. This results in longer-term, or decadal, variability in sea surface temperatures in the North Pacific and North Atlantic Oceans. Some scientists think that these changes in sea surface temperatures outside of the tropics also affect the strength and positions of winter storm tracks. Climatologists call these decadal modes of variability the "Pacific Decadal Oscillation" and the "North Atlantic Oscillation."

The North Atlantic Oscillation was discovered during the last century. Seamen traveling across the Atlantic Ocean noted that the winters tended to be mild in Greenland when they were harsh in Denmark, and vice versa. Today, we have a great deal more data to investigate such climate phenomena on a global scale. (To learn more about El Niño and predicting such phenomena, read Dr. Cane's essay in this section)

Centennial variability

Climate also varies over longer periods of time. About 200–400 years ago, glaciers advanced in Europe and northeastern America, a development referred to as the Little Ice Age. The extremely slow ocean currents in the deep ocean are thought to be connected to such centennial climate variability. It takes about 1,000 years for a water molecule to travel from the North Atlantic Ocean to the deep Pacific Ocean. There it might well up into the surface layer, and then return via warm currents through Indonesia, around Africa, and back to the northern North Atlantic. The lack of accurate ocean data for periods preceding the twentieth century makes it very hard to study such phenomena in great detail. Most of what we

know comes from several sources: changes in the thickness and chemical composition of layers deposited annually on large ice sheets and recorded in ice core samples, changes in the thickness and density of annual tree rings, and changes in sediment deposits in lakes or in the ocean. (Paul A. Mayewski's essay in this section describes how ice core samples record past climate.) However, these measured parameters reflect responses to changes in sea surface temperature, precipitation or winds, as opposed to the actual changes, so they must be interpreted with great care.

Ice Age variability

Climate change also occurs on time scales of periods 10,000 years or longer, referred to as glacial periods. Twenty thousand years ago, at the peak of the last big ice age, the oceans were dramatically different. For example, the sea level during this time was 150–300 meters below its current level, which allowed people to walk from Asia to America over the Aleutian Island chain. The Mediterranean Sea was a lake. Atmospheric and oceanic circulations were probably quite different, too. Enormous masses of water were accumulating on the land in the form of large ice shields, which might have increased the frequency and severity of storms. Scientists are just beginning to understand these past events. They discovered that just as the last ice age had terminated, a brief return back to glacial conditions occurred. This rapid change back to cold conditions (called the Younger Dryas; see Paul A. Mayewski's essay in this section) occurred within a few decades. It demonstrates that the climate is dynamic and can respond quickly to slow changes in the intensity of the Sun's radiation or other climate-changing influences.   ◓

Predicting El Niño

Mark A. Cane

Introduction

It first dawned on me that we were ready to try forecasting El Niño on the July Fourth weekend of 1985. I was reading a paper that claimed to see a premonition of an upcoming El Niño in the variations of the amount of heat in the northwestern part of the tropical Pacific Ocean. I wasn't convinced, but I was primed to hear the message, because my understanding of El Niño assigned the starring role to tropical Pacific heat content.

Mark A. Cane is the G. Unger Vetlesen Professor of Earth and Climate Sciences at Columbia University's Lamont-Doherty Earth Observatory.

I had first become interested in El Niño after the terrible winter of 1977, when I was still a student at MIT. It seemed that the persistence of patterns bringing storms to the northeast was somehow related to the ongoing El Niño in the tropical Pacific. I had a background in both meteorology and tropical oceanography, and was less interested in how the bad weather was generated than in why the El Niño was there in the first place. In the early 1980s, my then-student Steve Zebiak and I developed the first numerical model capable of simulating El Niño. Our goal was not prediction, but understanding.

Not that we were uninterested in prediction. The desire to have foreknowledge of climate variations must be as old as agriculture, and as widespread. Maybe more so, since even hunter-gatherers would have good reason to wonder what sort of weather lay ahead. In our culture, the most ancient and famous climate forecaster is the Biblical Joseph, who foresaw seven years of feast followed by seven years of famine.

What is El Niño?

The most dramatic, most energetic, and best-defined pattern of climate variability is called ENSO, which stands for "El Niño Southern Oscillation." It is a global set of unusual climatic conditions, and is made up of two components: an oceanographic one, called El Niño, and an atmospheric one, the Southern Oscillation. On average there is an El Niño event about every four years, but the cycle is highly irregular. Sometimes there are only two years between events; sometimes almost a decade. There are great variations in amplitude, too. Although each episode has its own peculiarities, all follow the same general pattern. At an early stage, unusually warm surface waters are found in the central to eastern equatorial Pacific. An increase in convective activity in the atmosphere—more clouds and rain, which heats the atmosphere, which in turn makes air

move around—is associated with these warmer surface temperatures. At a certain stage, the trade winds, which normally flow westward, slacken. Next, the tropical Pacific Ocean from the international date line to the South American coast warms dramatically, which further disrupts the trade winds. Heavy rains fall in normally arid regions of Peru and Ecuador, while droughts are experienced in Australia and southern Africa, and unusual tropical cyclones occur in regions such as French Polynesia and Hawaii. Farther away, the Indian monsoon, the seasonal rains of northeast Brazil, and even the regional climates over much of east Asia, North America, and Africa can be disrupted. All of this was known in the early 1980s, but it was not at all clear which of these things were essential to drive an El Niño event, and which were incidental.

The 1997/98 El Niño was the most extreme in the 150 years of instrumental data, and perhaps the largest in the past 1,000 years. It is estimated to have caused over $34 billion in damages and the loss of over 24,000 lives. Equatorial waters from the South American coast to the dateline warmed by an average of 3°C, with the warming at the coast exceeding 6°C. It rained so much in normally arid Peru that a lake 160 square kilometers in area and three meters deep formed in the coastal desert. Because of the extreme drought in Indonesia, huge forest fires burned in the rainforest on the Indonesian island of Kalamantan, spreading a haze of acrid smoke over much of southeast Asia. People as far away as Singapore were forced to wear surgical masks to venture outside, and the airport, one of the world's busiest and most sophisticated, was often shut down by the poor visibility.

A Brief History

Historically, El Niño referred to a warming of the coastal waters off Ecuador and Peru, which killed fish and birds and caused catastrophic flooding in coastal areas. El Niño has been documented as far back as 1726. In fact, it appears that El Niño rainfall a century earlier made it possible for the Spanish Conquistadors to cross an otherwise impenetrable desert.

The atmospheric component of ENSO, the Southern Oscillation, is a more recent discovery. The major figure behind the early studies was Sir Gilbert Walker, the Director-General of Observatories in India. Walker assumed his post in 1904, shortly after the famine resulting from the monsoon failure in 1899 (an El Niño year). He set out to predict the monsoon fluctuations, an activity begun by his predecessors after the disastrous monsoon of 1877 (also an El Niño year). Walker was aware of work that indicated swings of

atmospheric pressure at sea level from South America to the Indian-Australian region and back over a period of several years. He named these swings the Southern Oscillation.

In the next thirty years, Walker added correlates of this primary manifestation of the Southern Oscillation from all over the globe. He found that periods of low Southern Oscillation Index (low atmospheric pressure in Tahiti and high atmospheric pressure in Darwin, Australia) are characterized by heavy rainfall in the central equatorial Pacific, drought in India, warm winters in southwestern Canada and cold ones in the southeastern U.S. No conceptual framework supported Walker's patterns; his methods were strictly empirical. Since the data record did not go back many years, many dismissed his findings as random associations

Figure 1A: These block diagrams illustrate wind patterns, water temperature and ocean currents for normal conditions and during an El Niño Event. 1A Normal Conditions. Strong trade winds push warm surface water west, piling up water in the western Pacific. This allows cold deep water to rise along the South American coast.

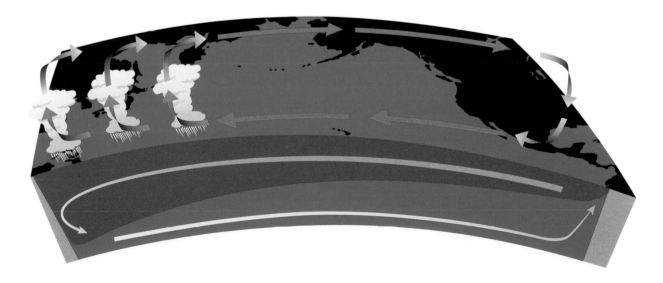

bound to show up in any short record. However, decades of new data have recently corroborated Walker's global correlations.

Oddly, Walker did not consider El Niño, and although both El Niño and the Southern Oscillation had been known at the turn of the century, it was only in the 1960s that the close connection between the two was finally appreciated, principally through the work of Jacob Bjerknes, a Norwegian then working at the University of California at Los Angeles. Bjerknes did more than point out the empirical relationship between the two; he also proposed an explanation that depends on a two-way coupling between the atmosphere and ocean. His ideas were prompted by observations of large-scale anomalies in the atmosphere and the tropical Pacific Ocean during 1957–58. A major

El Niño occurred in those years, bringing with it all the atmospheric changes connected to a low Southern Oscillation Index. A warming confined to coastal waters off South America could not cause global changes in the atmosphere, but the 1957 data showed that the rise in sea surface temperature extended across the Pacific. Ultimately involving a quarter of the circumference of the Earth, this warming was widespread enough to cause global atmospheric changes. Bjerknes suggested that this feature was common to all El Niño events, and he was correct. The term "El Niño" is now most often used to denote Pacific-wide oceanic changes.

Bjerknes suggested a tropical coupling between El Niño and the Southern Oscillation; he also proposed that the changes in heating

148 149

Figure 1B: During an El Niño Event the trade winds weaken, and warm water moves east, bringing heavy rainfall with it. Upwelling along the coast is blocked.

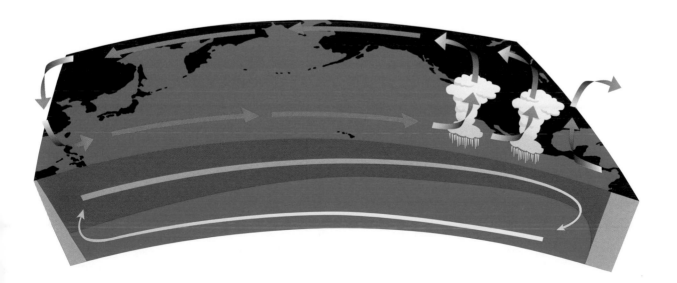

associated with tropical Pacific sea surface temperature anomalies drive changes in the global atmosphere. This idea is consistent with the global nature of Walker's Southern Oscillation. However, in his theory, the causes of ENSO are rooted solely in the coupling of the atmosphere and ocean in the tropical Pacific. They are entirely internal to the climate system, not responses to events such as volcanic eruptions, solar variations, or biological activity. While Bjerknes offered a persuasive explanation for both the normal and El Niño states, he stopped short of a hypothesis for the ENSO cycle, the perpetual change from a warm state to a cold one and back again.

An essential addition, equatorial ocean dynamics, was introduced by Klaus Wyrtki of the University of Hawaii in the 1970s. Wyrtki knew that the available data wasn't up to the task of telling us what the Pacific was doing, so he set up a network of island tide gauges to track the movements of tropical Pacific waters. His data showed that the sea surface temperature changes were not solely responses to changes in heating from the atmosphere above, but implicated ocean dynamics as well. Yet the theory still failed to account for the cycle. Work in the past two decades, especially that under the auspices of the international TOGA (Tropical Ocean Global Atmosphere) Programme, has provided theoretical and observational support for the Bjerknes-Wyrtki hypothesis.

From the beginning of our work on El Niño, Steve Zebiak and I conceived of our model as an embodiment of the essential physics singled out by Bjerknes and Wyrtki. Well, not quite from the beginning. There were a goodly number of false starts, as is usual in science. The model we developed was the first to give a successful simulation of the ENSO cycle. This numerical ENSO model depicts in a simplified manner the evolution of the tropical Pacific Ocean and overlying atmosphere. It is a dynamical model, rather than a statistical one; that is, it is built from the governing physical equations rather than from a sequence of past observations. Such dynamical models provide a means for physical interpretation and understanding of whatever they simulate. One of the most significant results of the model simulations was that ENSO recurred at irregular intervals as a result of strictly internal processes—in the absence of any imposed disturbances. Analysis of the model helped in developing a now widely-accepted theory that treats ENSO as an internal mode of oscillation of the coupled atmosphere-ocean system: a system that moves back and forth independent of external forces. The cycle is perpetuated by a continuous imbalance between the tightly coupled surface winds and sea surface temperatures on the one hand, and the ocean's more sluggish subsurface heat reservoir on the other. The movements of the layer of warm water can't keep up with the changing winds, so even as the winds coming out of an El Niño event are going back to normal, the Pacific is still sloshing around like an enormous bathtub, overshooting the normal position and driving the system toward a cold state. The cold state, now often called La Niña, is in many ways the opposite of El Niño in its effects. For example, while El Niño years tend to demonstrate above average spring/summer rainfall in the American Midwest, La Niña years are usually hot and dry. Thus, the ENSO cycle continues without end.

Predicting ENSO

This theory has some notable implications for the prediction of El Niño events. First, since the essential interactions take place in the tropical Pacific, data from that region alone may be sufficient for forecasting. Second, the memory of the coupled system resides in the subsurface thermal structure of the ocean. In this instance, memory refers to the way the heat has been arranged in the ocean. The distribution of heat has a certain shape because it remembers all that the winds have done to it. Warm waters of the upper ocean are separated from the cold waters of the abyssal ocean by a thin region of rapid temperature change. This region is called the thermocline. Where the warm layer is thicker, the thermocline is lower, and vice versa. The depth and deformation of the thermocline in the tropical Pacific Ocean is the key element in forecasting El Niño.

Back at Lamont after the Fourth of July weekend, Steve, our programmer Sean Dolan, and I went to work. The foremost problem was that the observations we needed most, the thermocline depths, didn't exist at the time. We worked around this by using observations of surface winds to drive the ocean component of our model. The model then created currents, thermocline depths, and temperatures that served as initial conditions for forecasts. Each forecast consisted of choosing the conditions corresponding to a particular time, and running the coupled model ahead to predict the evolution of the combined ocean-atmosphere system. By making "predictions" based on past periods, we could compare these retrospective forecasts directly with reality.

We began with July 1982, a time when the largest El Niño of the century was already well underway. We wanted to give our first try at forecasting the best possible chance. Besides, we knew that in September of 1982, experts had reached a consensus that there was no El Niño, so even recognizing one underway would be a small victory.

This first forecast was a success: the model evolved an amplified warming that peaked at the end of the year. It is the only thing I have done in my scientific career that worked the first time, despite some bugs in the code. Over

the next weeks we went on to try forecasts at three-month intervals for the entire period from 1970 to mid-1985. We felt we couldn't go earlier because the necessary wind data coverage was too spotty. And, of course, we couldn't go later because there was no data against which to compare our results. The results clearly demonstrated that the model could predict events at least one year ahead.

We wrote up the results and sent the paper off to *Science*. It was rejected. One reason was the journal's severe limits on length, which made it impossible to provide enough detail to satisfy a cautious reader. The broader reason the paper was rejected was a widespread and not unreasonable skepticism that El Niño could be predicted at all. Some thought that each El Niño was a separate event triggered by some random, unpredictable change in tropical weather. Our theory argues that ENSO is a system that evolves under fixed rules, rather than a sequence of random events. This idea had little currency in 1985. Moreover, even deterministic systems cannot necessarily be predicted terribly far ahead. If systems are what we call chaotic, which weather is, small differences in their starting points can quickly lead to large differences in where they end up. In other words, although the rules are clear, the outcome is less so. Here the situation was made worse by a lack of detailed historical records for vast regions of the tropical Pacific.

If, as had been convincingly demonstrated earlier, the chaotic nature of the atmosphere ensured that weather could not be predicted more than two weeks ahead, how could anyone claim to predict anything in the climate system a full year ahead? The answer has two parts. First, the tropics are more predictable than the temperate latitudes. Second, we are not trying to predict weather at a particular place and time, but average conditions over a large area and a few months' duration. Nothing known in 1985 precluded the possibility of doing so. On the other hand, nothing guaranteed that it could be done, and the claim to have done it came out of the blue, with little prior work to make it plausible.

Meanwhile, we began to notice that all the forecasts by our model initiated in 1985 agreed in calling for an El Niño event of moderate size at the end of 1986. Our retrospective forecasts from Januarys had been especially skillful, so when the January 1986 forecast confirmed the 1985 results, we began to consider what to do. Should we keep it within the academic community? After all, it could be wrong. The 1982/1983 El Niño was still a vivid, unhappy memory in Peru, Australia, California, and other hard-hit places. Warning that another was on its way, even if it was expected to be much smaller, would surely cause alarm. But it could be right—in fact, our research said it most likely was right. Shouldn't we warn people? If so, how?

First, we conferred with others at Lamont. Their judgment was that the science was sound, and that it was proper to make it public. We shortly sent off an account to *Nature* of the retrospective forecasts and the actual forecast for the upcoming January. We also held a press conference.

Making the forecast public did not receive universal approval from our colleagues in oceanography and meteorology. One of the more positive comments from a colleague was, "I thought it was great. Either you would be right, in which case it would look like we all knew what we were doing, or you would be wrong, in which case we could have a good laugh at your expense."

The eastern equatorial Pacific warmed through the first few months of 1986, in keeping with the expected El Niño. Just as though it had

been waiting for us to say something, the ocean began to cool shortly after our March press conference, and it stayed cold through the summer. It was a tense time for us, but the August data brought us some relief: the ocean finally began to warm. Nature caught up with the forecast during the fall, and evolved rapidly into an unmistakable El Niño. The forecast counted as a success, although differences in timing and other details showed that the prediction scheme was far from perfect.

The model was generally successful in forecasting the major events (1972, '76, '82, '86 and '91) a year or more ahead. Our forecast that 1991 would be a normal year was a turning point, because so many had seen "unmistakable" signs that an El Niño was coming by the end of 1991. The El Niño did come in 1992, and our successful forecast brought widespread acceptance that ENSO forecasting was for real. On the negative side, the model wasn't good at predicting smaller fluctuations, some of which are enough to influence climate elsewhere on the globe; 1993 was a case in point. Finally, our forecast failed dismally in 1997, which turned out to be one of the largest events of all time. As baseball manager Casey Stengel put it, "I had many years that I was not so successful as a ballplayer, as it is a game of skill." At first we thought the failure might be due to something new in the present state of the climate system, something our rather simple model could not handle. However, recent experiments show that the prediction would have been successful if we had fed the model with better initial data.

The Impacts of ENSO Events

Because ENSO affects global climate so powerfully, it stands to reason that human activities dependant on climate will also be impacted: agriculture, water resources, health, fisheries, tourism, and so on. The payoff from

ENSO forecasting lies in the possibility of acting in ways that minimize the negative impacts of an ENSO event and exploit the positive ones. It is far from obvious, however, that this can be done at this point in time. Even as ENSO forecasts become much better, the links to global climate will remain uncertain. Even where the links to climate are fairly certain, the impact on human activities such as agriculture is difficult to assess. There is real information in ENSO forecasts, but can we use it to the benefit of human society? Figuring out how to do that is the challenge ahead of us. ⦿

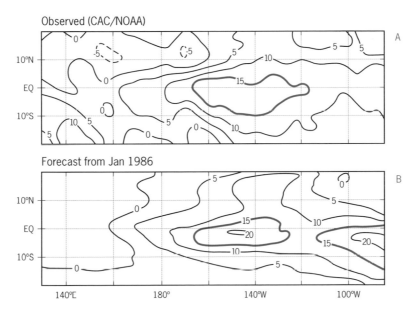

Sea Surface Temperature Anomalies (0.1C) January 1987

Figure 2:

A. Observed sea surface temperature anomalies for January 1987.

B. Sea surface temperature anomalies predicted by the Lamont atmosphere ocean model for the same month. These predictions were made one year in advance. Note the general similarity in the shape of the temperature anomaly counters, particularly the one labeled 15.

Global Warming

Charles F. Keller

Introduction

Global warming is in the news. While scientists agree that temperatures are rising, they disagree as to the causes and the rate of change. How much will temperatures rise, and how soon, and what will be the effects? There's a lot we still don't understand, because climate is enormously complicated. So are the factors that make the Earth habitable, of which temperature is only one. For example, certain types of air pollution cool the atmosphere and thus might act as agents to offset global warming, but they also make the air hard to breathe. Because

Charles F. Keller is the Center Director of the Los Alamos National Laboratory branch of the University of California's Institute of Geophysics and Planetary Physics.

climate change is so complex, scientists are using all kinds of scientific data and models to try to figure out what's actually happening.

In 1957, noted climatologist Roger Ravelle described the possibility of global warming from burning of fossil fuels. "Human beings are now carrying out a large-scale geophysical experiment of a kind that could not have happened in the past nor be reproduced in the future. Within a few centuries we are returning to the atmosphere and oceans the concentrated organic carbon stored in sedimentary rocks over hundreds of millions of years…"

Ravelle also encouraged climatologist Charles Keeling to measure carbon dioxide levels in the atmosphere. The resulting so-called Keeling Curve dramatically illustrates the rise of CO_2 above pre-industrial levels, a rise confirmed by measurements of the gases trapped in the thick layers of ice built up over thousands of years on Greenland and Antarctica (Figure 1). Perhaps more than anything else, this documented

atmospheric increase in CO_2, a powerful greenhouse gas (GHG), has served to bring home the possibility of humans warming the climate. The rise in CO_2 is largely due to the burning of fossil fuels. In fact, this rise makes warming a certainty since the physics is well understood and GHG warming is an obser-vational fact. (For more information on the carbon cycle, read Rachel Oxburgh's essay, "Earth: The Goldilocks Planet," in Section Six)

A Word About the So-called "Greenhouse Effect"

The Sun is hot, and shines most of its light in short-wavelength visible radiation (violet, blue, green, yellow, orange, and red). The Earth responds by absorbing some of this visible radiation, heating up, and irradiating energy back to space in the form of longer-wavelength, invisible infrared rays. Most of the gases in the

Figure 1: Atmospheric concentration of carbon dioxide compared with fossil fuel emissions. The graphs show the rise in greenhouse gases in Earth's atmosphere since the industrial revolution in the mid-1800s.

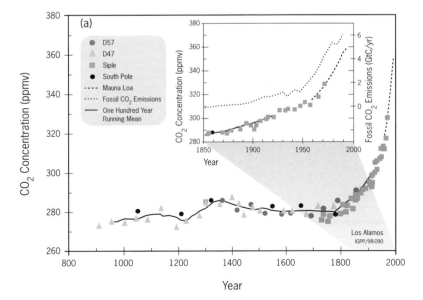

Earth's atmosphere allow both kinds of radiation to pass through relatively freely, but a few, called Greenhouse Gases (GHGs), pass the incoming visible radiation but absorb outgoing infrared radiation. This further heats the lower atmosphere and surface of the Earth, much the way sunlight "trapped" inside the glass of a florist's greenhouse warms the space within.

Water in the Atmosphere, or Why the Earth is Warm

The main natural greenhouse gases are water vapor, carbon dioxide, and methane. Water vapor is the most significant, and without it the average global temperature would be well below freezing. Humans have been adding increasing amounts of methane and CO_2 to the atmosphere over the past 100 years. When these human-produced greenhouse gases warm the planet, evaporation increases, and more and more water vapor is introduced into the atmosphere. Where in the atmosphere this

additional water vapor goes and what kinds of clouds it forms strongly influence how much additional warming is caused. These kinds of uncertainties make it hard to predict the effect of human-produced GHGs on global warming. Much about the relationship between global temperature and GHG levels remains unclear.

Climate in the Distant Past

The record of global temperatures for the past half million years can be derived from ice and sediment cores (Figure 2). Scientists use a variety of records to reconstruct past climate. Ice cores drilled from polar glaciers provide the most detailed record, in the form of layers of dust, chemicals, and gases which have been deposited with snow over hundreds of thousands of years. These layers reveal past climate characteristics, and many of their potential causes. (See Paul A. Mayewski's essay on the ice core record, in this section.)

The record clearly shows that our present warm climate is relatively rare. Most of the time the Earth likes to be much colder than it is now. The record of the past 11,000 years or so, the period of time in which civilization arose, shows

Figure 2: This graph shows the relationship between oxygen isotope levels measured from ice cores and global ice volume during the last 500,000 years.

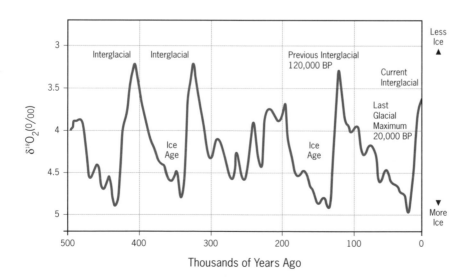

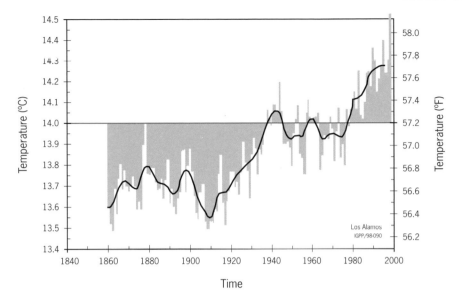

Figure 3: Measured global surface temperatures relative to the average since 1860. Note the dramatic rise in temperature during the 30-year period between 1961 and 1990.

that the Earth has been in a warm period, but also shows a slight but noticeable decrease in global temperatures. This is thought to be due largely to the combined effect of the wobble of the Earth's rotational axis and its elliptical orbit. The seasons are caused by the fact that the Earth's rotation axis is tilted with respect to the plane of its orbit around the Sun. Over thousands of years the direction that this rotation axis points changes. In addition, the Earth's orbit around the Sun is not a circle, but is slightly elliptical, causing its distance from the Sun to change during the year. Since the Northern Hemisphere (NH) has most of the land surface, it warms more for the same amount of sunlight than the ocean-dominated Southern Hemisphere. Thus, when NH summer occurs as the Earth is closest to the Sun, the climate is warmer, and 11,500 years later when NH summer occurs farthest from the Sun, the climate is cooler. Some 11,000 years ago, NH summer was closest to the Sun, but since then the rotation axis direction has been continuously changing until NH summer is occurring near the point in its orbit farthest from the Sun. Thus, the last 11,000 years have witnessed a small but steady decline in global temperatures. Recent

CO_2-induced warming is occurring on top of what might appear to have been a natural cooling. This is a complex interaction, and scientists don't understand it well enough to know how the combination will play out. (For more information on the orbital cycles of Earth, read the profile on Milutin Milankovitch in this section.)

The Strangely-Shaped Temperature Record of the Twentieth Century

The last hundred years have seen temperatures rise to levels not experienced on a global scale for over a thousand years (Figure 3). As predicted, the twentieth century rise generally coincides with the rise in human-produced GHGs in the atmosphere. However, while GHGs have increased at a steady rate, the actual temperature record fluctuates far more widely, as Figure 3 shows.

The warming can be divided into six parts:

1. cool temperatures until 1920
2. rapid warming from 1920–1940
3. slight cooling from 1940–1970
4. rapid warming from 1970–1985
5. slight warming from 1985–1997
6. large warming in 1998

If climate warmed simply in response to increases in GHG, the temperature increase would be smooth, and it is not. Climatologists have been working to explain this strangely-shaped temperature record.

GHGs, Aerosols, and Solar Activity

Because so many variables affect climate, it's difficult to establish a firm correlation between human activity (human produced GHG) and warming temperatures. The challenge is to establish whether a natural pattern exists, and then to determine what constitutes a departure from that pattern. Fortunately, the picture is becoming clearer. At least three separate computer simulations indicated not only that human-produced GHG had increased, but also that the effects of industrial aerosols had to be taken into account. (A byproduct of industrial pollution, these fine particles in the atmosphere actually cause cooling by scattering sunlight back into space.) These simulations have begun to explain the strangely-shaped temperature record. In particular, they have shown quantitatively why the global climate cooled between 1940 and 1974, when industrial air pollution was substantial, even as CO_2 levels steadily rose. In 1974, the U.S. Congress enacted the Clean Air Act, which resulted in a dramatic decrease in air pollution, at least by industry in the United States. Temperatures rose. For the first time, it was apparent that air pollution could scatter enough sunlight back into space to slow down the GHG warming, at least temporarily.

Computer simulations have become an extremely useful tool, but they still could not entirely account for the rapid temperature increase between 1920 and 1940. These have since been largely attributed to increases in solar activity in the form of sunspots.

While all three factors—GHGs, aerosols, and solar activity—have been operative to a lesser or greater extent over the past hundred or more years, the first rise (1920–1940) was largely due to increased solar activity, the leveling and slight decline (1940–1970) was caused by air pollution and perhaps a slight decline in solar activity, and the rise after 1975 is an increasingly clear signal of the effect of human-produced GHG. Between 1985 and 1997, however, temperatures have only risen slightly. This plateau effect is not fully understood, but increasing worldwide air pollution may be the reason; both the extremely rapid industrialization of the developing countries and the massive burning of forests produce light-scattering aerosols. In 1998, effects of a strong El Niño were linked to a large temperature increase, causing speculation that human GHGs were amplifying the lesser warming effect of the El Niño. (See Mark A. Cane's essay, in this section.)

Computer Models of Climate

Current computer models of the climate are becoming dramatically more accurate predictors of global warming. One example is the prediction of the global cooling and subsequent temperature rebound due to natural aerosols injected into the stratosphere by the eruption of Mount Pinatubo (a volcano in the Philippines) in 1991. These aerosols took four years to dissipate, during which time they reduced sunlight reaching the Earth's surface and lower atmosphere.

What the Models Predict for Warming in the Twenty-first Century

The latest computer models show that detailed prediction of warming in the twenty-first century due to human-produced GHGs will be difficult, because of the potentially large variations in aerosols and dust in the atmosphere, and unknown changes in solar activity. At the current rate of use of fossil fuel, it seems likely that CO_2 in the atmosphere could double, relative to pre-industrial levels, by the year 2050. Most computer models agree that such an increase could cause a temperature rise of some 2°C. The current uncertainties only heighten our concern, because while consequences might be fairly benign, they could also be disastrous to both humans and all other living things.

Potential Impacts of Global Warming

During the summer of 1999, the eastern half of the United States experienced weeks of above 90s temperatures combined with extremely high humidity and severe droughts. This led several states to ration both water and electrical energy (due to increased demand to run air conditioners). The number of people whose deaths could be attributed directly to this massive heat wave was over 200. This scenario is an example of what we might expect in the twenty-first century as human emissions of greenhouse gases cause additional global warming. While a rise in temperature of about 2°C may not seem great, that rise is an average over the entire Earth for an entire year. What we really expect is much larger departures from the normal temperatures in smaller regions of the Earth (half a continent) for shorter times (a month or less). These may be quite severe, and it is their effects that cause us concern. From such events, we expect problems such as: reduced crop yields due to droughts, extreme storms as the Earth attempts to "cool off," local outbreaks of insect infestation or insect-borne diseases such as malaria and dengue fever, and freak weather events such as massive ice storms rather than ordinary snowfall. Another example is sea level rise, which is predicted to be less than a meter. While that rise will be enough to cause problems with low-lying areas, another potentially larger impact may be that tidal surges, amplified by storms, will be much larger and more devastating as they breach natural dune barriers and cause destruction farther inland.

International Response

Recognizing that early detection of possible human-produced GHG warming of the Earth was an important but politically problematic issue, the United Nations and the World Meteorological Organization formed the Intergovernmental Panel on Climate Change (IPCC) in 1988. Made up of scientists from most nations, its task has been to assess the amount of potential warming. Scientists generally agree that organizations like the IPCC are the best way to achieve a responsible assessment of such emotionally charged issues.

The IPCC's first reports in 1990 indicated that humans were probably causing some global warming, but that nothing could be said for certain. But in its 1995 report, the IPCC changed its position ever so slightly and caused quite a stir. In the summary statement it wrote: "While significant uncertainties still remain… the balance of evidence suggests a discernible human influence on global climate."

Conclusion

The emerging picture of climate change, while complex, is increasingly comprehensible. Human activity (burning fossil fuel, changes in land use, air pollution, etc.) must be seen in relation to other factors—Earth's orbital changes, solar variability, and natural cycles— particularly in the oceans. If the stakes weren't so high, this would be one of the most fascinating scientific problems of our time, since it combines so many biological, chemical, and physical processes in a great, chaotic, and complex system. But if we are correct, the effect of human-produced GHGs is a recognizable factor in the observed warming of the global climate.

The ultimate benefit of the research described in this essay will be a better understanding of the possible climate of the near future. One fact is abundantly clear: the next quarter of a century will be a fascinating period in the study of climate change. ◗

An Ice Core Time Machine

Paul A. Mayewski

The Climate Challenge

Intense efforts are underway to determine the history and significance of human influences on climate, but our understanding of climate change is still hampered by how little we know about existing natural controls on climate. Mounting evidence points to a variety of natural forces that significantly affect climate, including variability in the amount of energy emitted by the Sun, planetary orbits, volcanic activity, ice dynamics, and changes in ocean circulation. These natural forces have produced dramatic—and often rapid—climate change.

Paul A. Mayewski is the Director of the Climate Studies Center and a Professor in the Institute for Quarternary Studies and Department of Geological Sciences at the University of Maine.

Our ability to understand climate change, to decipher the influence of human activity, and to predict future climate, lies in investigating both past and modern climate and in comparing the two. Because of the combination of natural and manmade factors at work during the past century, understanding and predicting modern climate is a particularly complex challenge.

Why Ice?

We think of snow and ice as ephemeral and temporary. Yet they can create and preserve a remarkable, long-term record of certain environmental changes. Snow accumulates wherever it is cold enough, primarily a function of latitude or elevation. As long as some snow remains to form a base for a new layer, snow will accumulate from year to year to form a glacier. As progressive layers of snow accumulate, various factors combine to preserve the ice more easily. The ice grows thicker, protecting the lower layers. The elevation at which it forms gets higher, and therefore colder. With ice cover, the originally dark land surface becomes light, which reflects incoming radiation and further cools the surface. All these conditions help preserve the snow and ice, which may remain in a frozen, unaltered state, for hundreds of thousands of years or more. Ice cores are obtained by drilling into the ice. The ice core record is interpreted from the core. The thickest ice is in Antarctica (approximately 4,800 meters thick with an average thickness of 2,200 meters). The longest ice core records available, on the order of several million years, are preserved in this 4,800-meter section.

When snow falls, it brings with it gases, dust, and dissolved chemicals from the atmosphere. Although scientists had already studied the composition and other characteristics of polar ice, it was not until the late 1970s that they began considering its properties as an environmental container. This new field of ice core research looks at various properties of ice to understand the chemistry of the atmosphere and nature of the climate at the time the snow accumulated to form the ice. Such information allows scientists to deduce how and in some cases why climate has changed in the past.

Using Ice Cores to Investigate Past Climate and Environment

Scientists use records from a variety of sources—such as instrumental observations, historical documents, deep-sea and continental sediments, tree rings, and ice cores—to reconstruct past environmental conditions reliably. Of these, ice core records recovered from polar glaciers provide the most direct and detailed information; each is a veritable "time machine" for viewing climates of the past.

The principles of this "time machine" are based on reconstructing a variety of climate characteristics, such as temperature, precipitation, wind speed, the place of origin of major air masses, and the direction these air masses moved over the polar region. These characteristics are revealed by examining the dust and gases that are deposited with snow as it accumulates on glacier surfaces, as well as by looking at the chemical composition of the ice itself. For example, the presence of methane in gas bubbles in ice indicates the productivity of biological systems, such as the extent of swamps in coastal regions, and whether or not there was a lot of free water on the surface to support these systems. Nitrous oxide and carbon monoxide levels in gas bubbles indicate how much combustion of fossil fuel was occurring at the time the gases were trapped in the accumulating snow. Sulfate comes from fossil fuel burnings and volcanic activity. Big spikes in ammonium indicate forest fires like the Tunguska fire that occurred in Siberia in 1908 and may have been caused by a meteorite impact.

Precise dating techniques make it possible to define the annual, and in some cases seasonal, layers of accumulation. Once recovered, these frozen pieces of atmosphere are analyzed for over fifty properties. Ice possesses different physical as well as chemical characteristics. The size of the crystals, the temperature, and the density of ice all change seasonally and over time. The electrical conductivity of ice indicates whether it is acidic, a preliminary indicator of volcanic activity. These data combine to reveal not only changes in the composition of the atmosphere and in the climate, but also many potential causes of these changes.

Because it is exceptionally detailed and long-lived, the ice core record has been attracting international attention. Much has focused on the deep ice cores recovered from Summit, Greenland and from Vostok in East Antarctica (Figure 1). The new perspective they provide may help answer some of the great scientific questions of our time: how does the climate system operate, and how has the composition of the atmosphere changed over time?

Figure 1: Location map for deep drilling sites: GISP2 (central Greenland), Vostok and Taylor Dome (East Antarctica) and Siple Dome (West Antarctica).

The Greenhouse Gas/Temperature Relationship

The climate records from the Vostok ice core, from top to bottom, make up the longest available continuous record of Antarctic climate. The oldest ice is approximately 420,000 years old. This record covers more than three full glacial/interglacial climate cycles (a full glacial/interglacial period is best described as beginning when a glacier forms and ending when it completely melts).

One of the most important discoveries to come from analysis of the Vostok ice core is the fact that a relationship exists between the concentration of atmospheric carbon dioxide and methane and the temperatures of Antarctic air during glacial cycles. Specifically, during glacial periods when temperatures were cooler, carbon dioxide and methane levels were, respectively, thirty percent and fifty percent lower than levels during interglacial periods, when global temperatures are warmer. These gases absorb a portion of the outgoing radiation and reflect it back to Earth, creating what is known as the greenhouse effect. (See Charles F. Keller's essay, in this section.) This means that with all other factors being equal, higher concentrations of greenhouse gases tend to raise global surface temperatures and lower concentrations allow surface temperatures to decrease. The strong correlation of carbon dioxide and methane with temperature suggest that lower temperatures (and the resulting greater continental ice volumes) of glacial periods were caused at least in part by lower carbon dioxide and methane levels. The correlation, which is evident in the climate record for at least the last 160,000 years, has been used to predict future climate changes related to the human-caused build-up of greenhouse gases. The ice cores demonstrate a strong relationship between changes in CO_2 and temperature, such that high

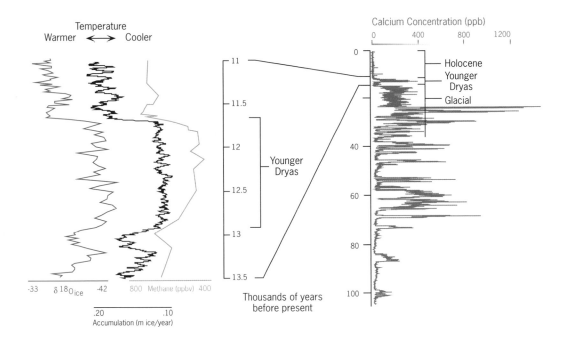

temperatures are coincident with increased levels of greenhouse gases.

Rapid Climate Change Events

In the past, the Earth's climate has experienced unprecedented swings that occur over decades—astonishingly fast in geologic terms—and persist for thousands of years. These rapid swings have now been recorded in two ice cores from central Greenland. This has dramatically contributed to our understanding of climate during the last glacial cycle, the past 100,000 years.

In 1993, the Greenland Ice Sheet Project Two (GISP2) successfully drilled to the base of the Greenland Ice Sheet. Along with its European companion project, the Greenland Ice Core Program (GRIP), the resulting record is the longest (dating back more than 250,000 years), high resolution, environmental record available from the Northern Hemisphere. The layers in the upper ninety percent of the GISP2 core match the layers in the GRIP core. This is important because it establishes the ice core record as reliable and consistent. The findings

Figure 2: This composite figure shows, on the right, the detailed calcium record from the GISP2 ice core for the past ~100,000 years, and shows the relative amount of dust (dust is calcium-rich) in the atmosphere over Greenland during that time. The changes in calcium concentrations document several abrupt, frequent, and massive changes in climate that characterize much of last 100,000 years (modified from Mayewski et al. 1994, 1997). On the left, the figure shows the Younger Dryas, the most recent of these abrupt changes that represented a sudden return to near glacial conditions about 13,000 years ago. The Younger Dryas was characterized by: cooler temperatures (15+/-3°C cooling relative to today); decreased accumulation rate; decreased CH_4; and increased atmospheric dust (see plot at right). The Younger Dryas lasted about 1,300 years. Data from Alley et al. (1993), Grootes et al. (1993), Brook et al. (1965), Severinghaus er al. (1998).

162 163

have since been further confirmed by findings in the upper ninety percent of cores from Vostok and Antarctica. The large-scale events recorded for the last 110,000 years in the two central Greenland ice cores unequivocally represent large climate deviations: massive reorganizations of the ocean-atmosphere system (Figure 2). They occur over decades or less and during periods in which ice age temperatures in central Greenland may have been as much as 20°C colder than today.

These events are of greatest magnitude during the glacial period, prior to 14,500 years ago.

The Most Recent Dramatic Climate Change: The Younger Dryas

Approximately 13,000 years ago, as ice from the last glaciation was rapidly receding, an event called the Younger Dryas occurred. This was a near return to glacial conditions marked by a number of climate changes (Figure 2): a drop in temperature of about 8–10°C in Greenland, a large decrease in atmospheric methane concentration, a more than tenfold increase in the amount of windblown dust and sea-salt in the atmosphere, and a twofold and greater increase in the rates of snow accumulation. These features signal cold, dry, and dusty conditions. High-resolution sampling over early and late stages of the Younger Dryas indicates that this event lasted 1,300 years, but began and ended in less than twenty—and perhaps in less than two—years.

The identification of such rapid climate change events in the methane record prompted scientists to seek out similar events in other regions. They looked in the marine records, in records such as windblown dust in Asia and tree rings that chronicle terrestrial events, and in ice cores from other parts of world. Rapid climate change events are documented in all these sites, although those in the North Atlantic have tended to be larger. What is not absolutely established is whether or not all of these events occurred at the same time, since no other records can be dated as precisely as the Greenland ice core. A discrepancy in timing could tell scientists a vast amount about the causes of the events.

Using Other Natural Records to Corroborate Ice Core Findings

The scientific process involves looking for records in more than one place. This cross-disciplinary approach enables scientists to check one set of findings against another for accuracy. Information from different sources also helps establish the scale, timing, and nature of past events. This in turn helps scientists to understand the possible causes of past climate change.

Ice core records provide a useful framework for interpreting records recovered from other natural archives. An important source of ancient climate records is North Atlantic marine sediment cores. These records also vary notably across thousand-year time spans, although the precise timing of events is not known. Marine sediment cores contain evidence of past climate change. One such signal consists of changes in the flux of ice-rafted debris—rock fragments carried by ice and deposited on the ocean floor when icebergs melt. The fragments, which are too large to be carried by ocean currents, record when icebergs moved farther out into the ocean than normal. Such massive discharges of ice from the land indicate immense instabilities in the shape of the ice. These discharges could be caused by several factors, including a rise in sea level, a rise in water temperature, or a dramatic increase in snow accumulation. Another climate signal is the abundance of foraminifera—tiny, hard-shelled marine organisms that are highly sensitive to temperature change. Particular shapes of foraminifera favor particular temperatures, and the compositions of their shells also depend on water temperature. Foraminifera indicate that during the last glaciation large pieces of glaciers were discharged into the North Atlantic at specific times. These discharges appear to correlate with rapid climate change events documented in the ice core record.

Rapid Climate Changes on a Global Scale

Evidence of rapid climate change events extends beyond the North Atlantic and polar regions. Marine cores from the Santa Barbara

Basin reveal disturbances in the ocean circulation patterns of the East Pacific region. Marine cores from the western Atlantic correlate with events in the Greenland ice core records as well. So does evidence in the form of debris from melting ice caps found in the North Pacific sediment. Abrupt changes in atmospheric circulation and precipitation patterns over eastern Asia that correlate in time with the rapid climate change events are documented by thick layers of wind-deposited clay (called loess) from central China. Records of ancient fluctuations in alpine glaciers, mountain snowlines, and Andean vegetation reveal climate fluctuations that also correspond with events recorded in the Greenland ice cores. All of these records combine to form a global picture of climate that has changed rapidly from time to time.

Scientists have yet to understand fully the timing and complex causes of these glacial-age climate fluctuations. However, evidence is building to reveal that some of these climate events are regularly timed. In addition, the cumulative effect on climate of many forces—such as changes in carbon dioxide, methane, water vapor, dust in the atmosphere (both natural and biogenic), and volcanism—can now be demonstrated. Cloud condensation nuclei cause water vapor to condense to form the small droplets that make up clouds. Clouds are important because they affect the radiation balance of Earth, by reflecting sunlight during the day and holding the heat in at night. The ice shows chemicals that indicate the relative presence of clouds.

Although the largest changes in climate occurred when there were big glaciers in the Northern Hemisphere, more subtle ones have occurred in the past 10,000 years. Research on the climate records for this time demonstrates that major changes in climate can occur over time periods significantly less than

that of a human lifespan. Scientists believe that certain rapid climate change events of the past 7,000–9,000 years have had a significant impact on human civilizations.

The Little Ice Age
The last one to two thousand years offer important opportunities for understanding the subtler variations that influence modern climate. It's generally agreed that glaciers around the world and the Arctic sea ice expanded during at least parts of the thirteenth to nineteenth centuries, a period called the Little Ice Age (LIA), and that warming occurred in some regions for several centuries prior to that, during what is called the Medieval Warm Period. The LIA had the most abrupt onset (1400–1430 A.D.) of any of the rapid climate change events of the past 10,000 years, and the onset may help us understand modern climate. It is our best analogy for a colder-than-present climate. Based on previous analogues and other rapid climate change events, the LIA should last longer than 600 years. Some scientists suggest that the climate state that was set up during the LIA may still be in existence, and that the rising temperatures over the past few decades mean that the natural climate system has been perturbed by humans.

The Influence of Human Activity on the Atmosphere
Humans have modified their environment since ancient times, but only since the beginning of the Industrial Revolution has their activity had a dramatic effect on a global scale. There's no doubt that it is affecting the composition of the atmosphere. Over the last 200 years, the world's population has increased by more than five hundred percent, and the amounts of carbon dioxide, nitrous oxide, and methane emissions have significantly increased as well (Figure 3A). The only continuous, season-to-season record of variability in CO_2 started in

The Anthropogenic Impact

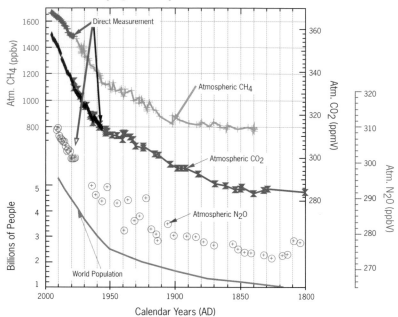

Figure 3A: This plot shows changes in the composition of the atmosphere over the last 200 years, determined from analysis of trapped gasses in Antarctic ice. The records show a rapid buildup of these human-generated (anthropogenic) gasses which relate to increasing world population (McEvedy and Jones, 1978). CH_4 is methane, CO_2 is carbon dioxide, and N_2O is nitrous oxide. Direct measurements taken since 1957 are also shown (Keeling et al., 1976). Data taken from Etheridge et al. (1992, 1996) for CO_2 and CH_4 and Machida et al. (1995) for N_2O.

1957, and it took several decades for people to realize that measurements were going straight up. The only preceding record of the atmosphere's composition is archived in the form of gases trapped in ice sheets. A reconstruction of the last 200 years shows a two-fold increase in methane, a twenty-five percent increase in carbon dioxide, and a ten percent increase in atmospheric nitrous oxide concentrations. These unprecedented levels are causing serious concern about the heat balance of the entire planet. Human activity also increases levels of sulfur aerosols, ozone, and

dust, to mention only a few emissions, which can either reinforce or counteract greenhouse gas effects on a local or regional scale. There's still some doubt as to the degree to which human activity is responsible, but no doubt that the increase in all three gases is directly related to the population explosion and accompanying industrial and land use stresses.

High-resolution analyses from a south Greenland ice core covering the last two centuries clearly demonstrate the difference between natural, pre-1900 levels of sulfate and nitrate, and twentieth century levels (Figure 3B). Concentrations have increased markedly since the turn of the century because of atmospheric pollution from North America and Eurasia. The sulfate record is detailed enough to document a decrease during the "Great Depression" of the 1930s, an increase due to renewed industrial activity that began during World War II, and the effects of the United States Clean Air Act of 1974.

The Future of Ice Core Research

Using ice cores and other paleoclimate tools, scientists are beginning to understand how the major features of the global climate system operate and how the chemistry of the atmosphere has changed over time periods of 100,000 years. The next challenge is to try to understand regional climate changes by looking at records of the last few thousand years in various localities. These records will be the best analogue for the way climate is today, because it's in the last 10,000 years that the geography of the planet—for example, the course of its rivers and distribution of vegetation—has been the same. Not every region can provide ice cores, but records from the Arctic, the Antarctic, and Nepal can be studied, and the data applied to other regions. Studying these ice cores and other climate records requires more and more detailed measurements, and increasingly sophisticated techniques for investigation. ☉

Figure 3B: This plot shows varying levels of sulfate (SO_4^{2-}, upper line) and nitrate (NO_3^-, lower line) with time. Measurements were made on the ice core from site 20D in southern Greenland. Examples of volcanic events recorded as spikes in the sulfate record are: 1. Laki (1783); 2. Tambora (1815); 3. Katmai (1912). Modified from Mayewski et al. (1990).

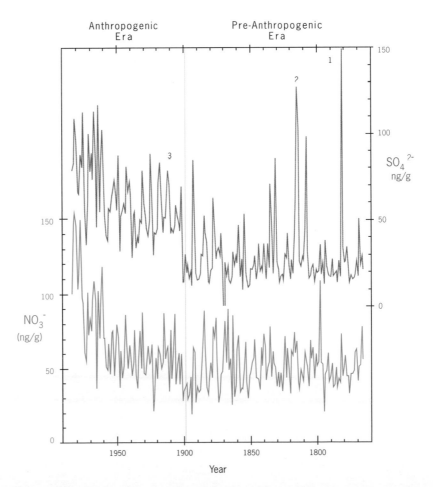

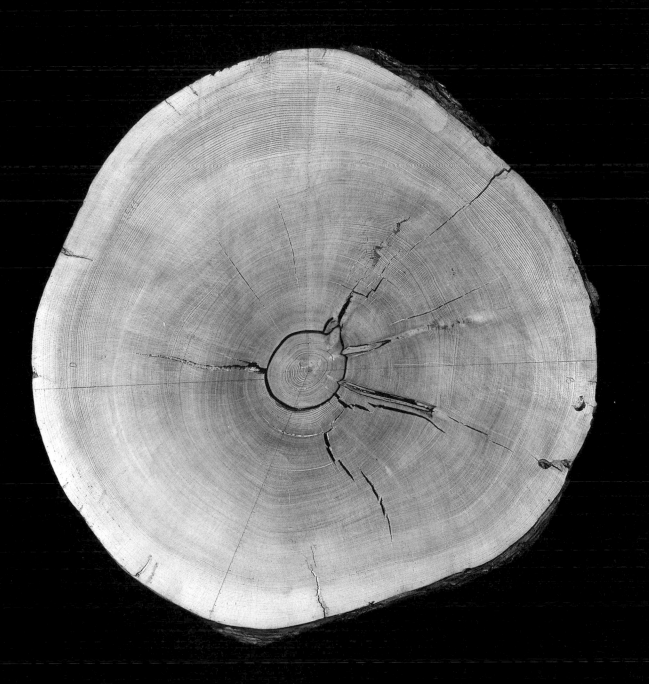

This tree ring section comes from Alaska's northern
treeline. It was cut from a white spruce that lived from
1441 to 1931.

Gordon Jacoby extracting a core from a white spruce tree near the treeline in Alaska. One can see the core sample on the extractor coming out of the tree between his two hands. Only a small hole is left in the tree. It fills with sap and heals in one or two growing seasons.

Studying Tree Rings to Learn About Global Climate

"Trees are an enormous feature in the global landscape," observes Gordon Jacoby. "They're fascinating in that they have the ability to record environmental changes." Tree rings generally grow wider during warm periods and narrower during cold ones, so their rate of growth provides a picture of Earth's temperature over the past centuries.

Co-founder of the Tree Ring Laboratory at Columbia University's Lamont-Doherty Earth Observatory, Jacoby is a dendroclimatologist. (Dendro is the Greek word for trees, and a climatologist is a scientist who studies climate.) Jacoby specializes in the study of annual growth patterns from old trees to see what they reflect about environmental conditions as they grew. "In most parts of the world, the written record of major environmental changes is short, so we

don't know the full range of variations that can occur naturally," he notes, "and thus we can't tell whether there have been recent, human-induced changes." From a network of tree ring data being collected around the entire Northern Hemisphere tree line, the Tree Ring Lab has reconstructed temperature records for the Arctic and Northern Hemisphere dating back to the 1600s.

Disturbance in the normal tree growth patterns can yield information about natural phenomena, such as earthquakes and volcanic eruptions, if the tree was located close enough to record the effects of a particular event and yet survive it. But the major application of tree ring research is dendroclimatology: studying the relationship between tree growth and global climate change. The field really took off in the 1970s, when computers were able to develop and handle complex statistical models from tree ring data.

A recent expedition took Jacoby, Lamont-Doherty colleague Rosanne D'Arrigo, and Tsevegyn Davaajarnts, a professor at the Institute of Botany at the Mongolian Academy of Sciences in Ulaan Bataar, to the Tarvagatay Mountains of western central Mongolia. There they took samples of 300- to 500-year old Siberian pines undisturbed by human activity and growing at the timberline more than 8,000 feet high, the northernmost limit of tree survival in that part of the world. "There the cold temperature limits trees, and growth is sensitive to temperature variations," explains Jacoby. To obtain the samples, Jacoby used something called "a Swedish increment borer, a cylindrical steel tube that can drill into a tree and extract a core five millimeters in diameter." He usually takes two cores per tree, each of which provides a complete cross-section of all of the annual rings from the center (the oldest rings) out to the bark. "It creates a small wound that usually heals within one or two growing seasons," says Jacoby, comparing the process to the biopsy performed by a doctor testing for disease.

Providing data from this part of the world for the first time, the Mongolian samples help scientists to better understand the fluctuations in Earth's recent temperature. Dating back to 1550, the record was revealing. "Three-hundred year annual temperature reconstructions for the Arctic and Northern Hemisphere, based on high-latitude tree ring data, indicate that the warming during the past century seen in instrumental data is unprecedented," wrote the three scientists in their paper Mongolian Tree Rings and 20th-century Warming. "Specifically, these general trends are cooler conditions (more narrow rings) in the early 1700s, followed by a warming (wider rings) for the mid- to late-1700s, abrupt cooling and continued cool conditions for much of the 1800s, and a warming trend for the late 1800s and much of the 1900s." The warming trends generally correspond with periods of increased solar brightness before the Industrial Revolution, and the cooler nineteenth century period with several major volcanic eruptions. (Volcanic particles block sunlight that would otherwise warm the atmosphere.) But sunlight alone doesn't account for the significant warming of recent decades.

"In any proxy you're not getting the actual temperature, you're getting something that responds to temperature, so you're getting a little biological noise in the temperature response," cautions Jacoby. Yet this pattern is strikingly similar to that found in trees elsewhere in the Northern Hemisphere, including Japan, North America, Scandinavia, and Russia, all of which confirm that relative to the past three hundred years, the twentieth century is unusually warm. Climatologists are working to understand the relationship between human factors, such as the buildup of industrial greenhouse gases in the atmosphere, and of natural variations such as sunspot or volcanic activity. "An objective is to understand exactly how the climate system works so one can begin anticipating the future possibilities," Jacoby says, "but we're really gun-shy about the word 'predict'."

"On the basis of the tree ring work alone, we can't tell which influences are caused by human activity," he elaborates. "But our information can be used by climate modelers and others, who will be able to detect whether there's some new element affecting the climate system. Tree rings say, 'this is what's happening.' Modelers say, 'this is unusual,' and look for the cause." New questions are constantly arising. For example, Jacoby is finding that in some parts of the northern tree line trees aren't responding to temperature change the way they did in the past. "Their relationship to the environment may have changed because it's been so warm for so long, or perhaps because of other factors we don't yet understand," he says. Jacoby is planning to continue using this tool in various regions where the records are sparse "to develop a better long-term record, meaning thousands of years. From those longer records we'll be able to assess the present situation more definitively."

As the tree ring record expands, scientists are finding out more about long-term temperature variations, extremes, and trends. For example, studies of moisture-sensitive trees are yielding information about the frequency of drought on a global scale. "We're looking in some tropical areas, such as Thailand and Indonesia, where few long-term records of climate exist, as a means of expanding the geographical extent of the science," says Jacoby. In tropical regions, where temperature isn't a limiting factor, dendroclimatologists look for a moisture-stress signal—a reduction in tree ring size due to lack of water. "In areas like eastern Indonesia, an El Niño event usually causes severe droughts," he explains, "so trees growing in that area can give you information about temperature change around the whole Pacific Ocean." ☻

Gordon Jacoby sampling a 300-to-500-year-old Siberian pine tree in the Tarvagatay Mountains of western Mongolia.

Hikers trekking on the Perito Moreno Glacier at Glacier National
Park in Patagonia, Argentina.

Milutin Milankovitch: Seeking the Cause of the Ice Ages: Profile

Portrait of Milutin Milankovitch (1879-1958) by
Paja Jovanovic, 1943.

What causes an ice age? Will another one occur? Since it had become generally recognized by the mid-nineteenth-century that much of Europe had once been covered by a great sheet of ice, scientists have been wondering what could cause such vast shifts in the Earth's climate. Some began looking for underlying astronomical causes. They already knew that the tilt of Earth's axis caused seasonal change, and also that small variations in Earth's orbit, over tens of thousands of years, affected the amount of solar energy reaching Earth. Several scientists had proposed the existence of a cycle of global winters, but none of their figures seemed accurate, and testing their reliability was difficult. Each theory was eventually shelved.

In 1911 a young Serbian mathematician, Milutin Milankovitch, decided to chart the ice ages of the Pleistocene. (The Pleistocene is the epoch that began 1.8 million years ago and ended about 11,500 years ago. It was characterized by lengthy ice ages, when glaciers covered large regions of the continents, interrupted by short interglacial periods, when the climate was temperate.) All Milankovitch's calculations were done by hand, and he worked at them obsessively for the next thirty years. He incorporated new information about small variations in the tilt of the Earth's axis, and factored in small orbital changes caused by the gravitational tug of other planets. Each of these orbital variations has its own time scale, and consequently they interact in different ways over time, but each one is regular. Going back 600,000 years in his computations, he carefully calculated the effect of these factors on incoming solar radiation across the Northern Hemisphere. The charts and tabulations

Milankovitch created are still used today. He also measured summer solar radiation curves in high northern latitudes, where the ice age glaciers originated, linking certain low points with four previous European Pleistocene ice ages. Ultimately, the mathematician arrived at a complete astronomical theory of glaciation.

The horizontal striations in this outcrop, in New York's Central Park, were caused by scouring from boulders imbedded beneath an advancing glacier.

On the basis of his analysis, Milankovitch concluded that Earth's orbit changes in three cycles of different lengths. The shape of Earth's orbit around the Sun changes from less to more and back to less elliptical in about 96,000 years. The Earth is tilted on its axis of rotation relative to the solar plane, currently at an angle of 23.5°. This tilt changes, however, from 21.5° to 24.5° and back again in about 41,000 years. Finally, the Earth's axis of spin wobbles with a period of 23,000 years. The challenges for Milankovitch were to understand when the three cycles were coincident with each other and how they worked together to influence insolation (the amount of solar radiation received by the Earth). Based on his computations, Milankovitch theorized variations of more than twenty percent in the amount of sunshine reaching the northern latitudes. In his 1941 account, Canon of Insolation and the Ice Age Problem, he suggested that this caused the waxing and waning of the great continental ice sheets.

Like that of several predecessors, Milankovitch's work was greeted with considerable excitement, but was then largely dismissed. Ice ages are difficult to date, partly because each erases much of the traces of its predecessor. However, the tables were turned by the late 1960s. Technical advances made it possible for geologists to study deep-sea sediment cores that contain a climate record going back millions of years. This climate record shows remarkably regular variations, which correlate with the mathematician's figures and which are now known as Milankovitch cycles. However, it is also clear that astronomical factors alone cannot cause the large changes that the Earth experienced. Other factors must also influence climate but scientists still do not know how. ☉

Section Six: **Why Is the Earth Habitable?**

Detail of a fragmented silver vein.

Introduction Edmond A. Mathez

Literally speaking, the Earth is habitable because it is the right distance from the Sun, and it has water and the six elements (carbon, oxygen, hydrogen, sulfur, phosphorus, and nitrogen) that comprise ninety-five percent of life. But habitability is in the eye of the beholder—my idea and your idea of habitability may not be the same as that of, say, a microbe. So, we could also ask, why the Earth is habitable for the more familiar life forms, ourselves and other mammals, for example. We could also ask, what made it possible for human civilization to develop. This section examines these questions.

Through the ages, the Earth's mean surface temperature has remained at a level that has allowed life to develop and evolve between the freezing and boiling points of water. In some scientific circles, ours has become known as the "Goldilocks planet"—with a surface not too hot and not too cold, but just right for life. Rachel Oxburgh describes why in her essay. Briefly, the Earth's surface temperature has remained constant because the amount of atmospheric CO_2 has remained approximately constant. CO_2 is an important greenhouse gas (see Charles F. Keller's essay on global warming in Section Five). The amount of CO_2 in the atmosphere (presently 0.35 percent) is controlled by the cycling of carbon through the various global "reservoirs". These are the atmosphere, the ocean (where carbon exists as the carbonate ion dissolved in the water), the solid Earth (where carbon is in the form of mainly limestone and carbonaceous shale) and, of course, the biosphere (all the carbon present in living organisms). Carbon dioxide from the atmosphere combines with water to form a weak acid, which reacts with rocks and minerals in the soil. During this weathering process, calcium, bicarbonate, and silicate are washed into the ocean. Marine organisms use these materials to make their shells. The shells pile up on the ocean floor and turn into sediment. The sediment turns to rock and is returned to the atmosphere and ocean by weathering and erosion. Alternatively, it gets subducted into the mantle, to be returned to the surface during volcanic eruptions.

Because of the rates at which carbon moves among the different reservoirs, it turns out that nearly all of it—99.5 percent to be exact—is trapped in rocks. A good way to appreciate how lucky this is for us is to consider Venus, one of our neighboring planets. Venus is not too different from the Earth in terms of bulk composition and size, and both have roughly the same amount of carbon. Instead of its carbon being locked up in rocks, however, on Venus nearly all is in the atmosphere, giving Venus a thick atmosphere composed almost entirely of CO_2. The atmosphere insulates the planet and results in a surface temperature of 460°C —too high to sustain life.

As I mentioned, habitability is in the eye of the beholder, and some microbes have a different view of this matter. Among them are the microbes that live in the rocks around the deep-sea hydrothermal vents, described in Deborah S. Kelley's essay. Driven by heat from magma bodies, ocean water circulates through the crust at the mid-ocean ridges. As it is drawn down through cracks in the rocks, the water is heated and reacts with the rocks, picking up metals and other compounds such as H_2S. When the hot, mineral-laden water is injected back into the cold ocean, minerals precipitate immediately from it to form massive chimney-like structures composed mostly of sulfides. This is, in fact, how some ore deposits form. As the hot water flows through them, these structures serve another purpose—they are incubators for microbes that live at high temperatures and eat the chemicals, toxic to us, supplied by the hot water. The microbes form the base of the food chain for a unique community of animals that live around the hydrothermal vents. This is one place on Earth where life lives not on the energy of the Sun, but on the chemical energy of the Earth itself. Some scientists believe that life began at these hydrothermal vents. The vents are where the Earth and the life sciences meet. The submarine hydrothermal vents are a source of some profound ideas, and they have captured the imaginations of many scientists. Among them is Veronique Robigou, the subject of one of this section's case studies.

The life around deep-sea hydrothermal vents has prompted us to think about where else life could exist—in other extreme environments on Earth and even on other planetary bodies. One of the scientists who studies life in the extremes of Earth—in this case, in the deep cold of Antarctica—and the possibility of life on other planets is Chris McKay, whose research is presented in another of this section's case studies.

The question of the origin of life is of course an important one. So the profile in this section is about Harold C. Urey, one of the most influential thinkers on this question, as well as other weighty matters, such as the origin of the planets and the climate on the Early Earth.

Hydrothermal vents may be fine for some microbes, but what about humans? Obviously, the Earth provides us with the resources that have allowed a society to develop—the metals, fossil fuels, water, and even seemingly mundane materials as sand and gravel used for concrete. The formation of metal deposits is probably mysterious to most people and many geologists alike, so the final essay in the section, by James Webster, describes how some deposits are formed.

We don't know if other planets contain life. But we have come to recognize that it exists and thrives on Earth due to a remarkable set of interactions happening all around us: in the delicate chemical balance of the atmosphere and oceans, in fluids seeping through the bedrock beneath our feet and in the oceanic crust, and ultimately in the ceaseless churnings of the inner Earth, where the power that shapes our planet resides. ❻

To explore the conditions that make the Earth habitable, I pose the following questions:

Why is the Earth habitable?

Rachel Oxburgh, a Lecturer in Geochemistry at Edinburgh University and Adjunct Associate Research Scientist at Columbia University's Lamont-Doherty Earth Observatory, discusses the carbon cycle, one factor that allows our planet to support life.

Can life survive without sunlight?

Deborah S. Kelley, an Assistant Professor in the Department of Oceanography at the University of Washington, describes the geology of black smoker sulfide chimneys, the chemistry of superheated seawater, and the life forms that flourish in these extreme environments.

Where do metals come from?

James Webster, the Chairman of the Division of Physical Sciences and a Curator in the Department of Earth and Planetary Sciences at the American Museum of Natural History, explains where metals, the key ingredient of the objects which surround us, originate and how they are concentrated in the Earth's crust.

Earth: The Goldilocks Planet

Rachel Oxburgh

One of the most remarkable features of our planet is that the temperature at its surface is not too hot, not too cold but just right for life to exist; Earth has been dubbed the "Goldilocks planet." If temperatures were to plummet much below 0°C (32°F), oceans, rivers, and lakes would freeze solid. Since plants and animals require liquid water for their metabolic functions, this would have drastic implications for life on Earth. Indeed, conditions such as these predominate on the barren surface of our neighbour planet Mars, where temperatures are -60°C (-76°F) and life, at least on its surface, does not exist. Conversely, if

Rachel Oxburgh is a Lecturer in Geochemistry at Edinburgh University and Adjunct Associate Research Scientist at Columbia University's Lamont-Doherty Earth Observatory.

temperatures were to soar above 100°C on Earth, all of the water would turn to steam. The absence of liquid water would lead to the annihilation of almost all life forms. Such furnace-like conditions prevail on our other planetary neighbour, Venus, where the average temperature is a roasting 460°C (about 900°F)—intolerable to all life. What is more, evidence from ancient rocks tells us that this "Goldilocks" state has been the norm for the past four-and-a-half-billon years. Geologists have found rocks spanning all of geologic time with features that show they must have been deposited in water. Thus, over this huge period of time, Earth has never frozen or boiled and has always maintained itself at a temperature that is "just right." The aim of this essay is to explain the machinery that controls our planet's temperature and how we think that the Earth, almost miraculously, has kept it tuned to near perfection for so long. But first, it is important to understand the factors that determine the temperature at the surface of the Earth.

Nearly all the energy that warms the surface of our planet comes from the Sun in the form of light. In order for the Earth to be kept at a habitable temperature, however, this light has to be converted to heat and the heat has to be trapped near the surface of the Earth. The conversion to heat happens when light hits surfaces that absorb it. Objects that absorb light look dark, and those that reflect it are light. Forests and oceans are good light absorbers while ice caps, snow, deserts, and clouds reflect light. Surprisingly enough, if the Earth's surface were one big sandy desert, it would be quite cold indeed since very little of the Sun's light would be absorbed and converted to heat. This is also why it is intolerable walking barefoot on a dark asphalt parking lot on a sunny day, while white sand beaches remain at a relatively more comfortable temperature.

The dark, absorbent surfaces on the Earth radiate energy in the form of infrared radiation back out to the atmosphere. However, if all of this energy were radiated straight back into space, it wouldn't do much to keep us warm. Luckily for us, greenhouse gases in the atmosphere absorb the infrared radiation and trap it in the atmosphere. Just as you would get cold in bed on a winter's night if you had no blanket, the Earth without its greenhouse insulation would have a temperature of -18°C. There are many greenhouse gases: water vapour, ozone, methane, chlorofluorocarbons, etc. However, the greenhouse gas that is thought to have exerted particular power over Earth's climate in the past is carbon dioxide. (See Charles F. Keller's essay, "Global Warming," to learn more about the role of CO_2 in controlling climate, in Section Five.) With no carbon dioxide in the atmosphere, the Earth would freeze over; with too much, temperatures everywhere would get too hot for life. However, it doesn't take much carbon dioxide to do this important job. Most of our atmosphere is nitrogen and oxygen, with CO_2 constituting only 0.35 percent. It may not surprise you, therefore, to find out that the reason Venus is so hot is that its atmosphere contains tens of thousands of times more CO_2 than Earth's. But in many other ways, Earth and Venus are not too different. So why has our Goldilocks planet always had an atmospheric CO_2 content that has been just right?

Only a tiny fraction (about 0.01 percent) of Earth's carbon resides in the atmosphere. The lion's share—on the order of 99.5 percent—is stored in carbonate rocks and sediments. Carbon dissolved in ocean water makes up about 0.4 percent of the world total, while fossil fuels (oil and gas), along with living and dead organic matter, amount to about 0.1 percent. To get an idea of the amounts of carbon in each

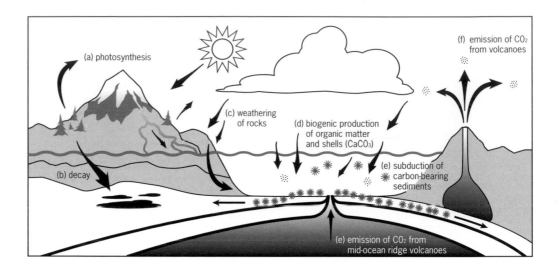

Figure 1: (a – b) short-term storage of carbon cycle (a) photosynthesis – respiration, (b) storage of carbon in organic materials in sedimentary rocks (oil, coal, natural gas, etc.), (c–g) long-term cycle involving (c) weathering of rocks, (d) the use of carbon from the ocean/atmosphere by organisms to produce organic matter and construct shells of calcium and carbonate ($CaCO_3$), (e) subduction of carbon-bearing sedimentary rocks in the Earth's mantle, (f) return of carbon to the atmosphere in the form of CO_2 released by volcanic processes at subduction zones, and (g) return of carbon in the form of CO_2 to the oceans by volcanic processes at mid-ocean ridges.

reservoir, imagine that the carbon in rocks and sediments is represented by a twenty-five-metre swimming pool full of water. On this scale, the carbon in the oceans would about fill a hot tub, fossil fuels would fill a bathtub, and atmospheric carbon would fill only a washbasin. The Earth's carbon reservoirs are rather like rest stops where a carbon atom resides for a while before moving on. These reservoirs contain carbon in different chemical forms and it can move between them in a variety of ways. The reservoirs and flows of carbon in the Earth make up the carbon cycle. Although the ocean and atmosphere are clearly different reservoirs, carbon dioxide mixes completely between the two quite quickly—in about 1,000 years. This might sound like a long time, but when we're talking about the billions of years over which

Earth's climate has been stable, it's no more than a blink of an eye. Thus on geologic time scales, the oceans and atmosphere behave as a single reservoir.

Now, we know that Earth's temperature has never been either boiling or sub-zero, so that means that the amount of carbon dioxide in the atmosphere can't have undergone huge and prolonged fluctuations. That's not to say it hasn't changed at all—in fact we know for sure that during the last ice age there was only about 0.2 percent carbon dioxide in the atmosphere, which is one of the reasons that the climate was so cold then (remember that it's about 0.35 percent today). While these changes have undoubtedly had an influence on climate, they have never been big enough to induce a long-term climate catastrophe. If you think back to the amounts of carbon in the various reservoirs, you'll see that even if all of the carbon in fossil fuels, plants, and trees were to be mixed into the ocean and atmosphere reservoir, it would only increase its size by about twenty-five percent. That's not true for the huge rocks and sediments stockpile. If just a small fraction of the carbon in that reservoir were to be transferred into the atmosphere, it would have a huge effect. (This is what happened on Venus—nearly all of the planet's carbon is in the atmosphere as CO_2.) So in order to understand why atmospheric carbon dioxide levels have stayed stable over billions of years, we need to look at the processes that move carbon between the rock-sediment and ocean-atmosphere reservoirs (Figure 1). The important processes are as follows:

Chemical weathering and carbonate sedimentation—the processes by which carbon is transferred from the ocean-atmosphere reservoir to the rock-sediment reservoir.

Carbon dioxide from the atmosphere dissolves in water at the Earth's surface to form a weak acid (carbonic acid—H_2CO_3). This acid reacts with rocks and minerals in soils, releasing the elements of which the rocks are made (e.g., calcium, sodium, silicon, etc.) into the water. During this reaction, the dissolved carbon dioxide is converted into bicarbonate ion (HCO_3^-). This process is known as chemical weathering. Rivers then carry the bicarbonate ions to the oceans where marine (ocean-dwelling) organisms use them to build their shells. When the organisms die, the shells fall to the seafloor and eventually become carbonate rocks and sediments. Thus by chemical weathering and carbonate sedimentation, carbon is removed from the ocean-atmosphere reservoir and transferred to the rock-sediment reservoir.

Volcanic eruptions and outgassing of the Earth—the processes that transfer carbon as carbon dioxide from the rocks and sediments in the Earth's interior back into the ocean-atmosphere reservoir

With time, the carbonate sediments that form on the seafloor are buried deeper and deeper into the Earth. The deeper in the Earth you go, the hotter it gets. When the sediments get deep and hot enough, chemical reactions occur that release the carbon as carbon dioxide. When volcanoes erupt, this carbon dioxide escapes back out from the deep Earth to the atmosphere.

So carbon enters the ocean-atmosphere reservoir from volcanoes and leaves it by chemical weathering of rocks and carbonate shell sedimentation in the oceans. Now the amount of carbon that is converted to bicarbonate ion during chemical weathering is always about the same as the amount buried in carbonate sediments on the seafloor. If chemical weathering begins to go faster for some reason, the rivers will carry more bicarbonate to the oceans. This carbon supply is greedily gobbled up by organisms to make their shells and the carbonate sediments from the shells of dead organisms pile up faster on the ocean floor. On the other hand, if chemical weathering rates become more sluggish, the supply of bicarbonate by rivers to the oceans dwindles. The organisms have less carbon to build their shells, so the shells pile up more slowly. What this shows you is that the flows of carbon from the atmosphere to the ocean (by chemical weathering) and from the ocean to the sediments (by shell formation) are always about the same—they are not independent of each other. So, the output from the ocean-atmosphere reservoir is really chemical weathering, since the amount of carbon transferred to the sediments on the seafloor is always the same as the supply by chemical weathering.

Let's get back to the Goldilocks question. If the amount of carbon in the ocean-atmosphere reservoir has stayed relatively constant over time, then the inputs and outputs to the ocean-atmosphere system must have remained in delicate balance. A tiny imbalance between volcanic activity and chemical weathering would lead to a climatic catastrophe. For instance, what would happen if more carbon were to be emitted by volcanoes than is taken up during chemical weathering? Levels of carbon dioxide in the atmosphere would build up until most of the Earth's carbon was in the ocean-atmosphere system. More and more of the Sun's energy would be trapped by Earth's greenhouse blanket (Figure 2) and we would be living (or, more likely, dying) in a Venus-like furnace. On the other hand, what if something causes chemical weathering to accelerate so that more CO_2 is sucked out of the atmosphere than is outgassed by volcanoes? Levels of carbon dioxide in the atmosphere would start to fall; before long there would be none left. Earth would become colder and colder until the oceans froze and we were living in an icehouse. It turns out that it would take about half a million years for these catastrophes to occur (again, a mere blink of an eye compared to the age of the Earth).

Now we know that the amount of CO_2 coming into the atmosphere from volcanoes and mid-ocean ridges has varied considerably over geologic time. Thus, the removal rate of carbon by chemical weathering must always have almost perfectly balanced the supply rate from volcanic activity. How can this have happened? One possibility is that changes in chemical weathering cause changes in volcanic outgassing so that the two match. However, it is difficult to imagine how the deep Earth could be aware of chemical weathering rates at the Earth's surface. The other possibility is that changes in volcanic outgassing lead to changes in chemical weathering so that the two maintain their balance. To explore this possibility, we need to understand the factors that control the rates of chemical weathering.

Nearly all chemical reactions go faster at higher temperatures. A rule of thumb is that for every 10°C increase in temperature, reaction rate

doubles. Thus the chemical weathering reactions between water and rocks begin to go faster as Earth gets warmer. How does this help solve the problem of the outgassing/weathering balance? If for some reason volcanic activity on Earth were to double, atmospheric CO_2 levels would start to escalate and the Earth would start to warm up. As temperatures rose, the chemical weathering reactions would start to go faster. More carbon dioxide would be sucked out of the atmosphere by weathering, shunted to the oceans in rivers as bicarbonate ion, and buried in ocean sediments as calcium carbonate shells. In this way, the extra carbon dioxide emitted by the volcanoes would be prevented from turning the planet into a furnace since it would be removed by increased chemical weathering. Thus we have an elegant means by which to explain Earth's remarkable thermostat.

Temperature isn't the only control on the rate at which chemical reactions occur. If a rock is ground up into grains, its surface area increases—each grain has an exposed surface that may react with the elements. The process by which rocks get broken up is called physical weathering. Much of this physical weathering occurs in the mountains. This is because mountain streams and rivers (which are fast-flowing because of steep slopes) carry rocks that knock against each other and are ground up. You will no doubt have seen piles of boulders and rocks at the bottom of mountains. Those piles are a result of the erosion of the mountains and the rocks that make them up will weather faster than if they had remained as the solid mountain. Since a new mountain range will increase the amount of physical weathering on Earth, it will increase the amount of chemical weathering as the rocks are being ground up. This will surely disrupt the carbon balance of a Goldilocks Earth.

Imagine if a new mountain range were to pop up in the middle of Kansas. You might expect that the physical weathering occurring in the new mountains would lead to faster chemical weathering. This would start sucking carbon dioxide out of the atmosphere faster than it was coming in from volcanoes. Earth's atmosphere would run out of CO_2, and Earth would turn into an icehouse in just a few hundred thousand years. Once more, however, Earth's clever thermostat would come to the rescue. Increased weathering rates lower CO_2 levels in the atmosphere. However, as CO_2 levels drop, so would the Earth's temperature, since its thinner greenhouse blanket traps less heat. This drop in temperature would cause chemical weathering reactions to slow down, decreasing the flow of carbon dioxide from the atmosphere. As long as removal of CO_2 by chemical weathering exceeds the supply by volcanoes, the Earth will continue to cool and as it does so, chemical weathering rates would continue to drop. At some point, the Earth would become cool enough that the chemical weathering rates slow down so that they exactly balance the input from volcanoes. Once more, Earth's thermostat saves us from disaster.

What you should now see is that our Earth's temperature is strongly dependent on the amount of carbon in the ocean-atmosphere system. A delicate balance between outgassing of volcanoes (input) and chemical weathering and carbonate sedimentation (output) controls the size of this reservoir. This balance is maintained by Earth's thermostat which makes chemical weathering accelerate when CO_2 builds up in the atmosphere and temperatures begin to rise, and turns it back down again if CO_2 gets low and the planet gets cool. In this way, the amount of carbon in the ocean-atmosphere system has remained relatively constant over the course of Earth history. Fortunately for us, our planet has stayed habitable to life. 							❻

Figure 2: Ultraviolet radiation from the Sun is absorbed by the surface (rocks, vegetation, water) and reemitted as infrared radiation. Ultraviolet radiation is reflected off clouds and snow unchanged. Infrared radiation absorbed by CO_2 molecules in the atmosphere is radiated back down to the Earth's surface. This radiation is what causes the greenhouse effect.

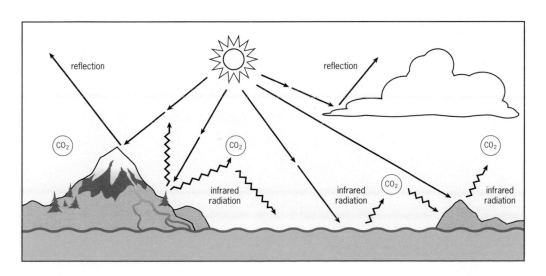

Black Smokers: Incubators on the Seafloor

Deborah S. Kelley

The Discovery of Black Smokers

In 1977 the scientific community was astonished by the discovery of hot springs on the ocean floor, thousands of meters below sea level. These sulfide chimneys or hot springs supported rich biological communities that thrived in the absence of sunlight. Equally surprising was the finding that these unusual animal colonies were sustained by microorganisms that feed on chemicals in hot water. The chemicals are released from magma

Deborah S. Kelley is Assistant Professor in the Department of Oceanography at the University of Washington.

In this illustration, the Alvin explores the base of the 15-story tall smoker dubbed Godzilla.

deep within the oceanic crust, then picked up by superheated seawater which circulates within the basaltic rocks that make up the ocean floor. The sulfide chimneys, which emit hot plumes laden with fine, dark particles of sulfide material became known as "black smokers." Sulfide is a term for a certain group of minerals that contain sulfur. The areas around black smokers are oases for vibrantly colored tube worm colonies, clams, crabs, and other animals in the desert of the surrounding deep-ocean environment.

Since that pivotal discovery, numerous underwater hot springs sites have been found along most of the major oceanic spreading centers (Figure 1A). Ongoing discoveries associated with these incredible environments continue to astound us. For example, the black smoker "Godzilla," discovered along the Endeavour Segment of the Juan de Fuca Ridge in 1991, was as high as a fifteen-story building and towered over the surrounding volcanic terrain before it toppled over in 1996 (Figure 2C). It is now also known that these "rocks" are literally alive with microbes that thrive within their warm, sulfide-rich, water-saturated interiors. Perhaps the most far-reaching idea to come from these hotsprings is that life itself may have originated within these dynamic systems, in which geological, chemical, and biological processes are intimately linked.

What Fuels the Black Smokers?

The spectacular development of vigorously venting black smokers on the ocean floor is fueled by circulation and heating of seawater at depths of 2–8 kilometers within the oceanic crust. The same process also fuels more subdued, lower-temperature venting systems. In volcanically active areas such as the East Pacific Rise (EPR, Figure 1A), the convection of heated seawater, or hydrothermal fluid, is driven by heating from a chamber of molten rock, or magma, at depths of two kilometers below the seafloor. The temperature of this magma is 1,200°C. In contrast, in areas such as the Endeavour Segment (Figure 1A) where there is little current volcanic activity, circulation is believed to be driven by the cracking of hot crystalline rocks, heated to 500–700°C, deep in the ocean crust. Along their downward journey from the ocean floor to depths of 2–8 kilometers beneath it, the fluids undergo significant changes in temperature and chemical composition as they approach the heat source. The fluids obtain their final chemical composition at the deepest point of this circulating process (Figure 1B).

Figure 1A: Global distribution of Black Smokers along the mid-ocean ridges. B. Cross-section of hydrothermal circulation. Cold seawater seeps into the seafloor, circulates near a heat source, becomes hot and buoyant, and flows into focused upwellings at hydrothermal fields.

A

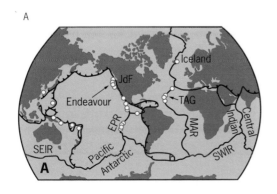

B

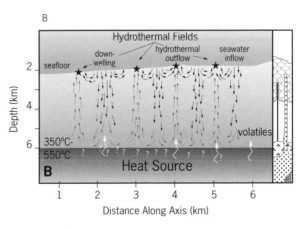

Special Seawater

The composition of the hydrothermal fluid, which is chemically modified, superheated seawater, is determined by three factors: the temperature of the rocks through which the fluids circulate; how much water has previously passed through that same crack network; and the composition of the rock. As cool, dense seawater migrates deep within the crust along the abundant large and fine-scale networks of cracks within these spreading environments, two important changes take place. First, the fluids interact and exchange elements with the surrounding host rock. Elements such as copper, zinc, iron, lead, sulfur, and silica are leached out of the rocks at temperatures of 350–550°C, and are incorporated in the hydrothermal fluid. Other elements such as sodium, magnesium, and calcium are added to the rock, modifying its original composition and mineralogy. In addition, gases such as hydrogen, methane, and carbon dioxide are added to the fluid when these compounds are directly released from magma chambers or leached from the enclosing host rock (Figure 1B). The compounds are critical nutrients for microbiological development at more shallow levels. Second, as the downwelling fluids approach the magma, their physical properties undergo dramatic changes. They become extremely buoyant, their viscosity and density decrease significantly and their ability to carry heat increases. Similar to the way water changes when it's heated in a pan, the now buoyant, metal-rich fluids rise up (through fractures in the seafloor), drawing in cooler fluids in their wake, and a circulation system develops (Figure 1B). This process is called convection.

How Black Smokers Form

The exact physical and chemical processes by which hydrothermal vents begin to develop are still poorly understood. In young hydrothermal systems, plumes of hot water that rise through the crust beneath the seafloor have to displace the surrounding cooler seawater saturating the shallow portions of the oceanic crust. In order for the high-temperature fluids to reach the seafloor, the channels through which this water rises must become progressively insulated by deposition of minerals. Once a venting system is established, however, the black smokers grow and evolve as the high temperature fluids vent onto the seafloor. The metal-rich fluids, heated to 350–400°C, mix turbulently with oxygen-rich, cold (2°C) seawater. This drastic temperature change causes solids to precipitate from the fluid. This process generates particles of sulfide minerals such as pyrite, chalcopyrite, and sphalerite, and sulfates such as anhydrite and barite.

Much of the fine-grained sulfide particles are carried upward into the buoyant, jet-like plumes that spew out of the vent at more than a meter per second; the particles are carried 100–200 meters up into the overlying ocean water, forming broad, extensive hydrothermal plumes. Some of these particles sink back to the ocean floor, oxidize and become sediment, and some are scavenged by microbial communities that live within the plumes. Sulfide and sulfate minerals that are not carried away in the hydrothermal plumes are deposited at the opening of the vent, causing the vent to grow upward over time. The chimney walls fracture periodically, which allows hot fluids to eject along the sides of the chimneys, and causes outward growth. These fluids may engulf and eventually fossilize tube worm colonies growing on the outer surfaces of the vents. In many oceanic systems, the high temperature (over 350–400°C), acidic, oxygen-poor, and sulfur- and metal-rich hydrothermal fluids boil as they rise up through the plumbing system beneath the black smokers. This boiling generates gas-rich vapors. In other, cooler systems, vents emit metal-rich water that may contain up to twice as much salt as the surrounding seawater. All of these processes come together at the seafloor in the formation of black smoker chimneys.

Different Formations Reflect Different Conditions

The spacing, growth rates, and mineralogical evolution of black smoker chimneys are complex and not well understood. Black smokers' shapes vary depending on different spreading environments, even when only a few hundred meters separate them (Figure 2). Along the mid-ocean ridge system, one of the largest single known deposits occurs on the Mid-Atlantic Ridge, a slow-spreading environment. The entire deposit, known as TAG (Figure 1A), is a large sulfide mound measuring 250 meters in diameter, and 50 meters high; it was probably formed by individual venting structures that combined into one deposit over time. Active venting is maintained by a black smoker

complex located on the top of the mound that hosts multiple, spire-shaped chimneys up to fifteen meters high, with fluid temperatures of around 370°C (700°F). White smokers on the margins of the mound discharge fluids at 265–300°C. Their venting fluids are white because the dark sulfide minerals precipitate within the mound before the fluids exit the chimney. Scientists think that in this area of active faulting, large intersecting faults have channeled flow to the hydrothermal mound episodically over a 50,000-year period.

In contrast to the large sulfide mounds that may typify deposits on the slow-spreading Mid-Atlantic Ridge, black smokers at the fast-spreading East Pacific Rise commonly occur as

Figure 2A: The Mothra Hydrothermal Fields host at least five sulfide chimney clusters, such as "Faulty Towers."

Figure 2B: The "Smoke and Mirrors" black smoker chimney.

Figure 2C: "Godzilla", a towering 15-story black smoker. Note that although each of these hydrothermal vents is located along the Juan de Fuca Ridge, their shape and structure vary greatly.

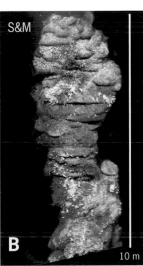

small, discrete individual structures rarely more than fifteen meters tall. Individual venting sites may be spaced 200–300 meters apart, and most venting occurs within shallow, fault-bounded central valleys, called grabens, which are located at the crest of the summit. In this volcanically active area, the small size of the deposits is probably due to at least two factors. First, molten basaltic rock frequently erupts and covers the hydrothermal vents and associated biological communities. Second, magma moves up from the magma chamber into the shallow underlying crust, which disturbs the fluid flow channels that had been established to feed hydrothermal vents. This causes the vents to shut down until the channels become reestablished, or even to relocate on a new site.

Somewhere in between these configurations is the Endeavour Segment of the Juan de Fuca Ridge, an extremely active hydrothermal area 300 kilometers off the coast of Washington State. It is one of the best-studied underwater environments. Active faulting (rather than volcanic activity) along this segment has produced a one kilometer wide, 100–200 meters deep graben in which four known vent fields are spaced 2–3 kilometers apart (Figure 1B). The fields, which are generally 400–500 meters long, host abundant large sulfide structures on top of which stand abundant black smokers. Areas where lower-temperature fluids vent more weakly are also common. In three of the fields, the most common sulfide forms include large, irregularly-shaped structures (up to eighteen meters tall) that host vigorously venting 350–400°C black smoker chimneys on their tops (Figure 2B). Lower-temperature venting of nutrient-rich hydrothermal fluids through the porous chimney walls support rich and diverse colonies of tube, sulfide, and palm worms, galatheid crabs, and a variety of snails and limpets (Figure 3A). On actively venting structures, these animal communities are so

dense that they obscure the underlying host rock. Many structures are characterized by stair-step arrays of large sulfide ledges that form an almost tree-like structure (Figure 2C). The most spectacular of these structures, "Godzilla" in the Endeavour Segment's High Rise Field, was forty-five meters high and contained 15–16 tiers of ledges up to seven meters in length which trapped 330°C fluids. This structure collapsed sometime in 1996.

In contrast to the steep mound and ledge shapes that typify most of the sulfide structures in the three other hydrothermal fields at Endeavour, the recently-discovered Mothra Hydrothermal Field hosts at least five sulfide clusters composed of multiple, isolated, and fused pinnacles that reach up to twenty-four meters in height (Figure 2A). Many structures are awash in cooler, slower-moving hydrothermal fluids (30–210°C) that support dense and diverse biological communities composed predominantly of a variety of worms, snails, and crabs. These animals colonize the steep outer surfaces of active chimneys. Venting fluids from some of these steep, isolated pinnacles reach 305°C. Horizontal fractures, marked by dense colonies of single-celled organisms that include filamentous bacteria and microbial mats, commonly cut the sulfide structures, and toppled structures are fairly common. Mothra is the largest hydrothermal field on the Endeavour, reaching over 500 meters in length.

Life in Extreme Environments
Within the walls of active smokers, where highly-specific combinations of thermal and chemical conditions exist, the limits to life are beginning to be defined. That is what makes the study of these and similar sulfide chimneys particularly exciting and important to underwater researchers (Figure 3A). Until twenty years ago, no one thought life would be

possible under these conditions. Many microorganisms recovered from these extreme environments belong to the domain Archaea, considered the most ancient of hyper-thermophilic organisms, those that thrive in 100°C water. In addition, scientists believe that the sulfide structures provide a window into processes similar to those operating deep within the seafloor, but which are not easily accessible with our current technologies. They are conducting intensive research into the linked mineralogical and biological interactions that take place within black smokers and the lower-temperature fluids around them.

The origin of hydrothermal systems is extremely ancient. For several related reasons, scientists theorize that life on the early Earth may have arisen in environments like these. One reason is that relentless collisions with large planetary bodies may have rendered both the shallow early ocean and the proto-continents uninhabitable. Furthermore, microbial life exists within hydrothermal systems that rely on chemosynthesis for energy in the absence of the sunlight. The study of black smokers is crucial to a broad range of microbiological and geological inquiries. Subsurface environments like these, both hot and oxygen-free, are believed to mimic most closely the conditions of Earth's earliest subsurface. From these studies, it is clear that volcanically active planets that contain liquid water may harbor life. Black smokers can thus be viewed as natural laboratories for studying microbial metabolisms thought to be ancient on Earth—and perhaps elsewhere. ☻

Figure 3A: Highly specific combinations of thermal and chemical conditions support dense communities of diverse life forms on the sulfide chimneys. B. Cross-section showing internal structure of the sulfide chimney. C. Nutrient-rich hydrothermal fluids vent through the porous chimney walls supporting colonies of tube, sulfide, and palm worms, Galatheid crabs and a variety of snails and limpets.

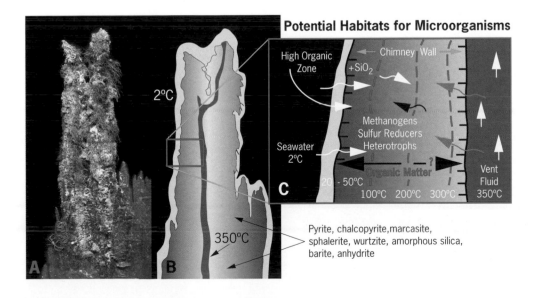

Potential Habitats for Microorganisms

High Organic Zone

Chimney Wall

+SiO₂

Methanogens
Sulfur Reducers
Heterotrophs

Seawater
2°C

Organic Matter

Vent
Fluid

2°C

350°C

20 - 50°C

100°C 200°C 300°C 350°C

C

A   B

Pyrite, chalcopyrite, marcasite, sphalerite, wurtzite, amorphous silica, barite, anhydrite

Resources of the Earth: Where Do Metals Come From?

James Webster

Humans rely on many resources other than air and water for modern civilization. Everything else we need must be either grown or mined. We extract and use great quantities of fossil fuels like coal, oil, and natural gas; materials like sand, gravel, and building stones for construction; and nonmetals like salt, sulfur, and diamonds for various industrial applications. All of these come from the Earth. Metals are also key resources, and they, too, come from the Earth.

James Webster is the Chairman of the Division of Physical Sciences and Curator in the Department of Earth and Planetary Sciences at the American Museum of Natural History.

The rocks and minerals of the Earth provide a large number of metals. In fact, metals comprise more than three-fourths of the 92 naturally-occuring elements. Of all the resources that we regularly employ, metals are unique in the many ways that they affect our daily lives, from the components in electronic devices such as computers to the drainpipes in our homes and under our city streets. Metals are also integral to the cars, buses, or trains that transport us to school or work most every day.

Many metals exhibit properties that make them extremely useful. The ten most heavily utilized metals are iron, aluminum, copper, gold, nickel, platinum, tin, lead, zinc, and molybdenum (listed in approximate decreasing order of worldwide usage). Iron is a dense and strong metal that is used in constructing steel structures like buildings and ships. Nickel and molybdenum are mixed with iron to increase the strength of some steel alloys. Copper and aluminum are in great demand as good conductors of electricity and heat, respectively, so cookware and electrical wiring are made from them. Platinum is used in catalytic converters in automobiles, and is also useful in the chemical industry, because it's relatively nonreactive when exposed to strong chemicals. Metals like gold, silver, and lead are malleable—they can be readily shaped by hammering or other means. Gold and silver are also valued because of their ductility, which means that they can be easily drawn out to form fine wire. Because of this and also because of their great beauty, they're used for jewelry and other decorative purposes. Zinc and tin are melted together or with other metals to form various alloys like brass or bronze. Metals are part of almost everything in our everyday life.

The principal question we want to address here is the source of these enormously useful materials. Where do metals come from? The answer, surprisingly, is stars. Metals were generated at extreme temperatures and pressures in the cores of massive stars billions of years ago. Through fusion reactions, lighter elements, like oxygen and silicon, combined with one another to make heavy elements like iron and nickel. The creation of other metals, like molybdenum and tin, involved high levels of nuclear radiation and processes that are far more complex. Later, the metals were ejected from stars by the solar winds and distributed in space by these solar winds and by the catastrophic explosions known as supernovae that cause the deaths of stars. Before the birth of the planets themselves, the metals combined with nonmetallic elements and other metals to form the great variety of minerals and rocks. These minerals and rocks were distributed within our inner solar system. They formed local agglomerations in places that coincide roughly with where the inner, rocky planets—Mercury, Venus, Earth, and Mars—are now found.

Metal Deposits and Mining

A metal deposit is a quantity of rock that contains a greater than average concentration, called an enrichment, in one or more metals. Most metal deposits on Earth are concentrated in and on the crust, which is the thin, outermost "layer" of our planet.

Before a mining company will expend the effort and money necessary to locate, drill, blast, and excavate metal-rich rocks from the Earth, it wants to make sure that a significant concentration of the metal of interest exists. That location must contain a greater than average concentration of metals in order for the metal to be extracted at a profit. Many economic factors influence the profitability of mining, including the demand for the metal at that time, the depth of the deposit from the surface, and the inherent expenses involved in extracting the metal from a rock.

The amount of metal that crustal rocks must contain in order to make mining profitable varies from metal to metal, depending on their abundance and value. Iron and aluminum, for example, are fairly abundant in the Earth's crust, and deposits of these metals need be enriched by only about four to five times their average concentration in crustal rocks to be mined at a profit. Platinum is one of the least abundant metals in the crust, but it is extremely valuable. In order for a platinum deposit to be profitable, the metal must be enriched by 500 to 600 times. Others, like molybdenum, gold, silver, and tin are some of the least abundant metals in the crust but are generally of less value, so their average concentration in rocks must be enriched by 3,000 to 4,000 times before they can be mined at a profit.

How are Metals Concentrated in the Earth's Crust?

The fluids that are found in rocks throughout the Earth's crust are the key agents of this natural process. Metal deposits are formed when these fluids extract trace concentrations—very small amounts—of metals from the rock. This involves several steps. First, a fluid of some type has to come into contact with a rock that contains metal. Second, these fluids must remain in contact with the rock long enough for the metals to dissolve into the fluid. Heat

Figure 1: Jim Webster with the head geologist of the Climax mine, Mr. Reese Ganster (standing), in the mine, studying a large, molybdenum-rich, granite boulder that was chosen for display in the Gottesman Hall of Planet Earth at the American Museum of Natural History.

accelerates this process. Third, the fluid has to transport the dissolved metals by moving through interconnected spaces in the rock such as fractures or cracks, or perhaps spaces between the grains, in order to transport the metals. This journey is a slow one. Fourth, the fluid undergoes one of two kinds of changes. The change is thermal if the fluid moves into cooler rock and its temperature lowers. Or it could undergo a change in chemical composition if it comes into contact with chemically reactive rocks. Either of these changes destabilizes the dissolved metals, and they form a solid mineral, or crystallize. Metals tend to form minerals on a solid surfaces, and therefore crystallize along the surface of fractures or in other empty spaces within the rock.

What are these fluids? They are naturally-occurring liquids like hot salty water, molten sulfur-rich liquids, molten silicate rock—known as magma—or even cold seawater. The Earth exhibits a great diversity of metal deposits, because rocks contain a large number of metals and because many types of fluids may be involved in their transport and concentration. As they flow, the fluids continue to dissolve the rock and actually change its mineralogy. It's an ongoing, complex set of interactions, because each type of rock, metal, and fluid reacts differently to each of the other components. Some of these fluids react more than others, and the extent of the reaction also depends on the rocks through which they flow. For example, the molten sulfides associated with silicate magmas dissolve platinum and related metals, whereas cold seawater has little effect on it. Of the tenmetals mentioned at the beginning of this essay, all except aluminum, platinum, and nickel are mined mostly from deposits that were formed by the actions of hot, watery fluids—hydrothermal fluids—flowing through rocks.

Case Study of Ore Formation: The Climax Molybdenum Mine, Colorado
The largest known deposit of molybdenum metal on Earth is 125 kilometers southwest of Denver at Climax, Colorado (Figure 1). The Climax mine is located in the Mosquito Range of the Rocky Mountains at an elevation of 3,600 meters. Mining at Climax began in 1911, but this mine is no longer operating. The price of molybdenum dropped in the 1980s, and it became cheaper to obtain the metal from deposits elsewhere in the world.

The molybdenum fills a myriad of fine fractures within the Climax granites. Granites are igneous rocks that form as magma cools and crystals of the common silicate minerals quartz (made of silicon and oxygen) and feldspar (sodium, potassium, aluminum, silicon, and oxygen) form out of the melt.

As is typical when metal is concentrated by the actions of hydrothermal fluids, the process of depositing metal at Climax involved three ingredients: metal-bearing rocks, heat energy, and water. Here, the water for the fluids, the heat energy necessary to cause them to flow, and the molybdenum-bearing source rocks were provided by the molten granite magmas and hot, solidified granites. Between 33 and 18 million years ago, granite magmas rose within the crust of the Rocky Mountains to a final depth of roughly one-and-a-half kilometers below the surface. Because the magmas were intruded into colder rocks, they cooled and crystals of quartz and feldspar formed and grew. Crystallization began along the granites' outer borders, which cooled faster, while the cores of the magmas remained molten for a longer time. Magmas are composed of two parts: solid crystals and liquid melt. The liquid parts of the Climax magmas were gradually enriched in a variety of elements, including molybdenum, sulfur, water, and chlorine,

because quartz and feldspar do not incorporate these elements as they crystallize.

After extensive crystallization, the concentrations of water and chlorine in the molten granites became too high for the residual melt to contain them. This forced bubbles of hydrothermal fluid to form in the melt. The bubbles contained chlorine, an important component because scientists have determined through laboratory experiments that a little bit of chlorine in hot water goes a long way in dissolving and transporting metals. For example, hydrothermal fluids containing up to forty percent chlorine may remove as much as eighty percent of the molybdenum from a granite magma. This observation is important because fluids that contain the highest concentrations of chlorine are more likely to generate highly profitable metal deposits. How

do geologists know that the Climax fluids were enriched in chlorine? Figure 3 shows a photograph of microscopic inclusion in a crystal of quartz in a granite sample from Climax. An inclusion occurs when a foreign substance is trapped inside a mineral or rock. The inclusion, which appears roughly rectangular and is about 0.025 millimeters long and 0.015 millimeters wide, formed when hydrothermal fluid was trapped as a tiny bubble on the surface of a quartz grain as it grew in the magma. The inclusion contains a black spherical gas bubble, a large cubic crystal, and several other dark crystals; the remaining space is filled with salty water. The bubble and each of the crystals formed and grew in the inclusion while the granites were cooling, because hydrothermal fluids dissolve progressively lower and lower concentrations of solids and gases as temperature decreases. The large cubic crystal

Figure 2: A polished slab that was cut from the boulder (note the 45-centimeter rule shown for scale). Close examination of the slab shows fractures that are 1 to 5 mm wide. These fractures are filled with several minerals: molybdenite (made of molybdenum and sulfur), pyrite (iron and sulfur), and quartz. The slab, like the granites at Climax, is crosscut by four different sets of fractures. In the slab, one set is oriented vertically, one horizontally, and the other two sets cut diagonally across the vertical and horizontal fractures.

in the inclusion is a grain of salt that consists mainly of chlorine and sodium. Since an inclusion is a closed system—nothing was added and nothing removed after it formed within the host crystal—it without a doubt represents the composition of the hot fluid that was in the magma. The presence of the cubic chloride crystal therefore proves that the molybdenum-bearing hydrothermal fluid at Climax also contained high concentrations of chlorine.

After the fluid bubbles formed, they eventually dissolved molybdenum, sulfur, silicon, and iron from the granite magmas, the first step in the process of concentrating molybdenum. Slowly, the abundance and size of fluid bubbles increased and the bubbles built up sufficient pressure to crack and fracture the solidified outer parts of the Climax granites. Fluids rise toward the surface of the Earth through magma and through the networks of fractures. They rise because hot fluids are less dense than the surrounding magma or rocks. As the fluid bubbles rose through the cooler, solidified margins of the Climax granites, they reacted chemically with the rocks and cooled. When a fluid cools, it cannot contain as large a quantity of metals as when it is hot. The reduction in temperature and chemical reactions in the Climax granites caused molybdenum, iron, sulfur, and silicon to form crystals in the fractures (Figure 2). This was the final step involved in concentrating and depositing the molybdenum.

The metal-bearing veins and fractures of most other hydrothermal deposits contain numerous bands or layers of minerals. The bands in each

vein indicate that the fractures were held open for long periods of time by fluid pressure—to provide open space for the minerals to grow into—and that the composition of the fluid changed over time, depositing various bands of minerals along the walls of the fracture. The deposition of metals at Climax was different, however. It involved multiple episodes of fracturing the rocks and repeated filling of fractures by the deposition of minerals. Here, the fluid pressure built up and cracked the granites in one particular direction; molybdenite, pyrite, and quartz were deposited, and the rock fractures closed up because fluid pressure decreased as the fluid leaked away. Afterward, fluid pressure built up again, another set of fractures formed and cut the granites in a new direction, minerals were deposited in this set of fractures, and again fluid pressure decreased as this fluid leaked away. These processes were repeated at least twice, creating the rich molybdenum deposits which drew miners to Climax, Colorado. 🌐

Figure 3: A microscopic inclusion in a crystal of quartz in a granite sample from Climax. An inclusion occurs when a foreign substance is trapped inside a mineral or rock. The inclusion appears roughly rectangular and is about 0.025 mm long and 0.015 mm wide.

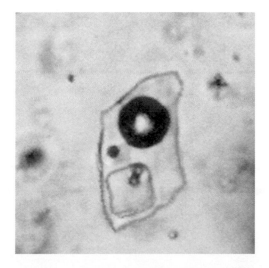

A global view of Mars.

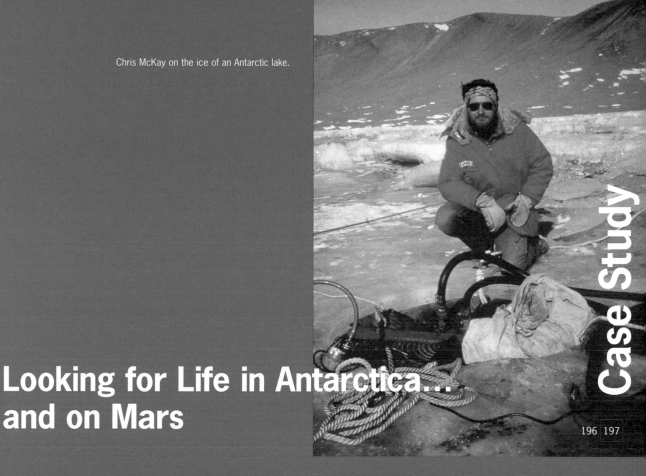

Chris McKay on the ice of an Antarctic lake.

Looking for Life in Antarctica... and on Mars

"The question I'm interested in is, 'Is there, or was there, life on other planets?'" says Chris McKay. A planetary scientist with the National Aeronautics and Space Administration, Dr. McKay is an astrobiologist: he studies the origin and distribution of life in the universe. He was a first-year graduate student when the Viking spacecraft landed on Mars in 1976, and "that did it," he recalls. "The Viking results seemed to suggest that the elements needed for life were present on Mars but there was no life there. That puzzle, that paradox, is what got me interested in the whole thing."

Satellite images show what might be long river channels on Mars, suggesting that liquid water once flowed on the surface of a much warmer planet. "The most interesting thing we've learned from all the missions to Mars, beginning with Mariner 6 in 1969, is that the planet had

an early Earth-like period," explains McKay. "It had water, it had active volcanism, and we believe it had a thicker atmosphere which would have kept the surface warmer. These dry riverbeds lie in cratered terrain more than 3.5 billion years old. Fossil remains show that microbial life on Earth existed at that time."

"Those Viking images from orbit are the decisive information," he says. They established the framework into which information from other sources fit. "I don't think we could have determined that early Mars was more Earth-like than today from meteorites directly," he says. "And I don't think anyone would have taken the meteorites seriously if we hadn't had the pictures. All that hullabaloo about finding life" (in 1996 scientists discovered that Martian meteorite ALH84001 contains tiny, egg-shaped structures which possibly constitute evidence of fossilized bacteria) "makes sense because of what we already knew."

"As planets go, Earth and Mars are still very similar bodies, although their evolutionary histories diverged several billion years ago," wrote McKay and co-authors E. Imre Friedmann and Michael Meyer in an article in *The Planetary Report* titled "From Siberia to Mars." "Geological processes seem to have obliterated evidence of life on Earth before 3.5 billion years ago; at least we haven't found any yet. But two-thirds of the Martian surface dates back further than that, so Mars may actually hold the best record of the events leading to the origin of life on Mars and on Earth, even if there is no life there today."

McKay's research takes place in the harshest climates on Earth, "because Mars is dry and cold," he explains cheerfully. The dry valleys of Antarctica are a prime destination because, with less than an inch of precipitation a year and an average temperature of -20°, they are the most Mars-like places on Earth. There, on windswept mountain slopes, tiny life forms—algae, lichens, bacteria—grow in the rocks. These microscopic organisms are called cryptoendoliths (from the Greek, crypto=hidden, endo=in, lith=rock). If Mars ever had life, biologists reason, its last survivors might have resembled these cryptoendoliths. And similar environments on Mars could have provided a refuge long after conditions on the surface became too cold and dry to support life.

The astrobiologist's main tool is a data logger, "a little computer that I leave out all year long to record basic environmental information: temperature light, humidity, wind." To monitor conditions within the rock, he drills little holes into which he inserts tiny moisture and temperature sensors that hook up to the data logger. The other invaluable tool is the microscope: "a portable one that I take into the field, plus a low-power hand lens, for looking at the organisms that live in these environments." McKay and his colleagues monitor a number of sites around the world, maintaining them for several years and sometimes longer. "For example, after four years in the Atacama desert in Chile, we still haven't seen any rain. We're waiting to see how long it takes."

Scientists also extract core samples from sites where life forms could have been preserved by the freezing temperatures. Seventy-five feet into the Siberian permafrost, they've found large numbers of 3.5 million-year-old-bacteria, which appear unharmed when thawed. Astrobiologists like Chris McKay hope that conditions at Mars's South Pole, parts of which have remained frozen at temperatures of -70°C since the creation of the planet, might have preserved similar tiny life forms. Any such organisms would have been killed by background radiation from ever-present radioactive elements in the Martian soil. However, the extreme cold would be ideal for preserving their intact bodies, which scientists could then subject to chemical and genetic analysis.

McKay's next trip will probably be back to the Atacama or to his station in Antarctica. But he's keeping a sharp eye on outer space, since NASA uses information from him and other astrobiologists to decide where else in the solar system to search for life. "Of particular interest to me in terms of biology is Europa, one of Jupiter's moons. We think that underneath the ice there's an ocean," he explains. "We have missions going to many of the planets, though more to Mars than all others put together." McKay thinks that the optimum destination would be the bottoms of dried Martian lakes. "These represent a place where life could have survived long after life outside the lake was dead, just as we see in dry valleys of

Antarctica," he points out. "They're also a good place for fossil remains, because they settle in calm lake bottoms. That's a big advantage over a riverbed." Because they're easily identifiable, ancient lakebeds also make a practical target—and a nice flat landing space—for a spacecraft. Nevertheless the technical obstacles—"getting to Mars, much less landing there"—are daunting. "It's a long way," admits the astrobiologist.

McKay thinks the current program of robotic exploration will eventually lead up to human exploration of Mars, "maybe in the first couple of decades of the next century." Landing in the permafrost of Mars's southern region and drilling down deep in this colder, less sunny environment would be even more difficult than

a touchdown on a Martian lakebed. But that's where the best chance exists of recovering actual organisms, dead or alive. "Distinguishing whether any such life forms are built of the same building blocks as life on Earth would be relatively easy," says McKay. "Now we can do Polymerase Chain Reaction, genetic sequencing, with exquisite accuracy. If it looks like us, that will be clear. But if it looks like a bunch of strange biomolecules forming the body of an alien microbe, we will have a hard time understanding how its biochemistry worked. I'd love to have that challenge." ❂

A ventifact, a wind-carved rock, from Antarctica.

Submersible *Alvin* being launched from the stern of the research vessel Atlantis II.

Veronique Robigou in the *Alvin* submersible on dive 3457 at 2,250 meters depth. On her most recent dive to the Godzilla sulfide structure in August 1999, Veronique observed and documented the growth of the spectacular sulfide edifice.

Mapping Hot Springs on the Deep Ocean Floor

"When I was ten, I knew I wanted to do science, and at twenty I knew the science would be geology," recounts Veronique Robigou with a smile. "It was only at the end of my twenties, purely by chance, that I fell into a group doing science related to black smokers." Black smokers are chimney-like structures made mainly of sulfides from which hot, mineral-laden waters vent. When these hot fluids enter the cold seawater, the minerals which are dissolved in the water precipitate out, giving the appearance of black smoke. The first black smokers weren't discovered until 1977, as scientists began to explore the deep ocean with submersibles. A student at the time, Robigou said to herself, "If these famous people just discovered these new environments, I can really be a scientist. I can make discoveries."

"The planet loses heat in many ways, and one of the ways is to release heat into the ocean at the volcanically active vents," explains Robigou. "What was unexpected was that some kind of life was going to be associated with these systems—and in such abundance and variety!" That life could exist at all under such extreme conditions came as a big surprise. At these depths, the water pressure is 220 bars compared to 1 bar at the surface of the ocean where human beings live and function. The chemical and thermal differences between the seawater, which is near freezing at 2°C, and the hydrothermal fluid that spews out of the chimneys (a superheated 350°C) are extreme. "It is completely dark. Yet the black smokers are filled with bizarre forms of life which somehow thrive in this harsh, sunless environment."

"I'm really interested in the rock," says Robigou, who is a marine geologist. "But the first thing that I see, before I can even see the rock, are the biological communities." These consist of many types of worms, along with limpets, mussels, sea spiders, many kinds of snails, and mats of filamentous bacteria. "Even after many dives, I am always amazed at the amount and variety of animals that live on top of each other, competing for the warm environment in which they find the nutrients to survive." It is this convergence of many systems—geological, thermal, chemical, and biological—which makes ocean vent systems so intriguing, and so instructive.

Robigou left her native France to learn field mapping at the University of California, Los Angeles. She thinks of mapping as "the first step in understanding the geological story. It establishes the frame you need before you can even start to ask questions." Robigou's mapping expertise put her in good stead when she began working with the University of Washington scientists who were studying the black smokers, but she also had to pioneer new techniques.

"It's frustrating not to be able to work the way you do on land," observes Robigou. At 2,250 meters below sea level, the scientists are confined to submersibles, "enclosed in a cabin of some sort, with a limited view." Lights illuminate only a small piece of terrain eight to ten meters in front of the submersible, "so I have to build in my mind an image of what these things actually look like from top to bottom." Sometimes instead of a submersible, the scientists use an ROV, or Remotely Operated Vehicle, which Robigou describes as "a robot that does the work while we stay on board the ship and survey the landscape from video monitors on board." The advantage is that during such dives, she can work with her whole team—typically several geologists, a pilot, a co-pilot, data logger, and a number of other scientists. The disadvantage, she points out, is that "you don't get a three-dimensional feel of the landscape. You are not immersed in the environment."

"The other important parameter is that when you walk on a trail in the California mountains, you have a map and you know where you are," she continues. "On the seafloor you don't always know exactly where you are, because for a long time we didn't have accurate, high-resolution maps." A big part of her work in the mid-1980s was to develop with other geologists, engineers, and pilots of submersibles faster ways to process navigational data from sound waves. These sound waves describe the shape of the landforms on the seafloor, and are plotted on maps on which Robigou can then overlay her geological observations.

In order to create detailed maps of the ocean floor, Robigou and her colleagues had to come up with new techniques for analyzing data. She relies principally on information from two sources: video images of the seafloor's basic features, and personal observations. "I describe everything I see through the porthole," explains the geologist. "In a submersible, one person looks to the right (starboard), the other to the left (port), and the camera faces the front. So in order to get a three-dimensional sense, you have to put all the information together."

In addition to mapping the seafloor, Robigou creates mosaic portraits of individual black smoker chimneys in order to track their evolution. "One of the attractive things about the black smoker environment is that we actually see the rocks forming in front of us," she observes. "In order to understand how all the mechanisms interact—the geological, geophysical, chemical, and biological processes—we first need to document things through time." She has been studying Godzilla, a sulfide structure about the size of a fifteen-story building, discovered in 1991 on the Juan de Fuca Ridge, 290 kilometers off the coast of Washington State. Robigou's first task was to document what she saw. "We analyzed video footage to create a detailed geological map, and also created a detailed scientific rendering of what Godzilla looks like, not just from the top but from the sides."

To document changes over time, the scientists go back once a year—or when funding is available—to re-image structures and see how they have changed. In 1996 Robigou went back to Godzilla and found a structure only twenty meters high. "I actually didn't recognize it. Then it dawned on me that if I looked at the base, I'd see big pieces of what used to be Godzilla." Sure enough, the colossal black smoker had fallen over, probably in an earthquake. "So the challenge, once I got over my shock, was to take advantage of the incredible opportunity to watch it grow again," she explains. "Over the last two years, a spire which was about one-and-a-half meters high has grown to ten meters. We don't yet know why they grow so fast. But we've realized that these systems recover their equilibrium very quickly, then grow at a slower rate."

An educator as well as a scientist, Robigou and her colleague Dr. John R. Delaney initiated a program in 1996 called "Research and Education: Volcanoes, Exploration and Life" (REVEL) which invites science teachers to participate in oceangoing expeditions. "They have to act and work like scientists, and they say it changes their lives, because they've been part of the real thing. The kids look at them in a different way, too," she continues. "They think, 'If the teacher can do it, so can I, because we're from the same community.'"

Robigou finds the black smokers a perfect educational tool. "All of the different sciences, which we usually learn in a separate way, come together. You can't make progress with the particular aspect which fascinates you without integrating what the other sciences are teaching you about the system," she explains. The same lesson applies on a larger scale. "You can really make the connection between this sulfide mineral growing at the bottom of the ocean, the emergence of new material from the ocean floor, and the convection of heat which involves the entire planet." The overarching message is clear: "There's still so much that we need to learn and understand. We are discovering new things all the time." ☻

202 | 203

Urey's student and colleague Stanley Miller in their lab at the University of Chicago in 1953. In their experiments, they attempted to study the origins of life by duplicating the conditions which would have existed on the primordial Earth. Their apparatus contained a "sea" of purified, sterile water under an "atmosphere" of hydrogen, methane, and ammonia. Incandescent electrodes imitated lighting and ultraviolet radiation from the Sun. After a week or so the result was a tar-like residue containing amino acids, the basic building blocks of life.

Harold C. Urey: Profile

Discoverer of Deuterium and Investigator of the Origin of Life, the Origin of the Planets, and the Climate of the Early Earth

Typically referred to as a "Nobel Prize-winning chemist," Harold Urey did indeed make invaluable contributions to our understanding of the atom in general and isotopes in particular. But from the very beginning he was drawn to other disciplines. Urey's Ph.D. in chemistry was preceded by a Bachelor of Science degree in zoology from the University of Montana and was followed by a year in Copenhagen at the Institute for Theoretical Physics, run by the renowned physicist Niels Bohr. Urey "settled down" to teach chemistry, but he continued to invade the traditional domain of physical scientists. The result was a remarkable body of work, not just in the field of physical chemistry, but also in geochemistry, lunar science, and astrochemistry.

In 1929, Urey became a professor of chemistry at Columbia University and wrote *Atoms, Molecules, and Quanta* (with A. E. Ruark) that

was published the next year. It was there that he tackled the problem of "heavy hydrogen." Other chemists had suggested that there might be a form of hydrogen atom with twice the mass of the ordinary hydrogen atom—a heavy isotope—although if it existed at all, it would only be in a small concentration. (An element is identified by the number of protons in its atomic nucleus. But different nuclei of the same element can have a different number of neutrons. Isotopes are atoms with the same number of protons but with a different number of neutrons.) Urey reasoned that if liquid hydrogen were slowly evaporated, most of the heavy hydrogen would remain in the liquid residue. He was right, and eventually Urey produced a liquid residue that contained enough of the heavy isotope to be detected through a spectroscope. The isotope was named deuterium, and in 1934 Urey received the Nobel Prize in chemistry for his discovery. Refusing to travel to Sweden because his wife was pregnant, he delivered his Nobel lecture the following year. Urey went on to devise a large-scale process for obtaining water containing high proportions of deuterium— so-called "heavy water."

Urey next applied his energies to separating isotopes of other elements. Deducing that heavy isotopes would have a slightly slower reaction time than their lighter twins, he devised ways in which to build up those differences into measurable quantities. By the late 1930s, he was able to create high concentrations of isotopes such as carbon-13 and nitrogen-15, found in minute quantities in natural carbon and nitrogen. The chemist's work was interrupted by World War II, in which he played a critical role. As director of the Atomic Bomb Project at Columbia University, Urey directed the production of isotopes of boron, hydrogen, and uranium. From uranium-238 he separated the rare isotope uranium-235, essential to the development of the atom bomb. After the war, hydrogen-2 (the deuterium Urey had discovered) was used to make the even more destructive H-bomb.

Urey was deeply concerned with the danger posed by the nuclear weapons he had helped create, and became an extremely active advocate of nuclear arms control. In 1945 he took a position as professor of chemistry at the University of Chicago and turned to an entirely new area of study: geochemistry. Once again, his knowledge of isotopes was put to ingenious use, this time in calculating the temperature of ancient oceans. Because isotopes react more slowly at colder temperatures, the properties of certain isotopes found in fossil shells reflect the temperature of the ocean at the time the shells were formed. Using oxygen isotopes as a "paleothermometer," Urey and his co-workers analyzed tiny fossil shells from ocean sediment cores to create a history of changing ocean temperatures over long geologic periods.

It was in 1953, while he was still at the University of Chicago, that Urey and graduate student Stanley L. Miller performed a landmark experiment that brought the problem of the origin of life into the laboratory. Complex organic molecules called amino acids are the building blocks of the proteins necessary for life. At the time, they had been found only in living systems. Urey believed that life was common in the universe, and that these building blocks must have spontaneously come into being on the early Earth. He and Miller set out to demonstrate how this could have happened. They filled a flask with methane, hydrogen, ammonia, and steam, which was thought to replicate the early atmosphere, and passed 60,000-volt electric charges through it to simulate lightning. Miller sat by the crackling apparatus for a week, then analyzed the chemicals in the water. They were full of amino acids. Most scientists now believe that the early atmosphere had a different composition than the one tested by Miller and Urey, and that organic molecules originated by other mechanisms. Yet the experiment galvanized the scientific community to think about how life may have begun.

Urey also studied the chemical make-up of the Sun, Moon and planets, and formulated detailed theories about the origin of the solar system. He believed that planets were built up by the accretion of smaller, mainly metallic fragments at relatively low temperatures, and that the Moon was formed separately. His second book was called The Planets.

A man of deeply-held political convictions, Urey took a courageous early stand against Senator Joseph McCarthy's anti-Communist campaign. He remained an anti-war activist, and opposed the use of nuclear power. Urey's legacy is a significant one. The isotope-labeling techniques he introduced have proven immensely valuable for researchers in many fields. He was widely published on many subjects, receiving countless awards and honorary degrees. He died in 1981, at the age of 88. ❻

About the Gottesman Hall of Planet Earth

A view of the central element of the Hall, an internally projected hemisphere that shows Earth from space.

On 12 June, 1999, the American Museum of Natural History opened the Gottesman Hall of Planet Earth (HoPE), a new permanent exhibition that tells the story of modern Earth science.

The central challenge in creating this hall was to devise a narrative that accurately represented the breadth and complexity of the science and yet fully engaged audiences of diverse ages and backgrounds. The exhibition team solved this problem by organizing the exhibit around a series of five key questions that span current scientific knowledge:

● How has the Earth evolved?

● How do we "read" rocks?

● Why are there ocean basins, continents, and mountains?

● What causes climate and climate change?

● Why is the Earth habitable?

In keeping with the Museum's long tradition of representing the natural world through artifacts and specimens, the answers to these five questions are explored largely through samples, the evidence of Earth's dynamic processes. Large, dramatic, and touchable samples are the focal points of each section, and they capture the visitor's imagination, just as they capture the imagination of scientists who study them in the field. These large samples, many of which are described in more detail below, also speak to the scales of geologic time and space in a concrete and vivid way.

The Earth Hemisphere and Earth Event Wall

A unique globe occupies the center of the Hall. Rather than a traditional sphere, this globe is an internally projected hemisphere, approximately 2.6 meters across and mounted on the ceiling. The hemisphere serves as a screen that shows

THREE TYPES OF ROCK

re are three kinds of rock: igneous, sedi- | accumulate in layers. Metamorphic rocks
tary, and metamorphic. Igneous rocks | result when existing rocks are changed by
n when molten rock (magma or lava) cools | heat, pressure, or reactive fluids, such as hot,
olidifies. Sedimentary rocks originate | mineral-laden water. Most rocks are made of
 particles settle out of water or air, or | minerals containing silicon and oxygen, the
ecipitation of minerals from water. They | most abundant elements in the Earth's crust.

An exposed magma chamber

Wind-formed cross-beds
in Navajo sandstone

Folded marble, or metamorphosed
limestone

These samples of the three rock types are mounted in front of a recreation of a cliff at Siccar Point, Scotland.

the Earth as we might imagine it while sailing through space above the planet. The image is a simulation based on NOAA and US Air Force weather satellite data: as the Earth slowly rotates, the clouds disappear; then the vegetation and ice are stripped away; and finally, the oceans are drained, revealing a rocky planet without water, atmosphere, or life. The globe appears to glow in the low ambient light and hangs above a small amphitheater, giving the area a contemplative ambience.

Adjacent to the globe is the Earth Event Wall (EEW), a projected map of the Earth that provides a global context for current events, such as volcanic eruptions, earthquakes, and large storms. Embedded in the map are windows that appear in sequence to tell brief stories with video and headlines; below the map, interactive computers offer visitors the opportunity to select stories and learn about them in more detail. The EEW demonstrates that the Earth is dynamic; keeps the exhibition current; and provides a direct means of communication between the scientific community and museum visitors. Over the first six months of its operation, more than twenty stories were produced, including accounts of the ongoing activity of Montserrat volcano; the effect of the 1998 hurricane Mitch on Honduras; the destructive tornadoes that struck Oklahoma City on May 3, 1999; and the August 17, 1999 Izmit earthquake in Turkey. The stories are archived, and most are accessible at http://earthbulletin.amnh.org.

Samples

A total of 168 samples are displayed in the hall, many of them raised on modernistic steel pedestals. Most have both rough and polished faces exposed. Organized around the key questions of the exhibit, these samples tell the Earth's story.

These impressive specimens of igneous rock are from Glass Mountain, on Medicine Lake Volcano in California.

How has the Earth evolved? Among the striking samples in this section is a 3075-kilogram boulder of banded iron formation from near Cobalt, Ontario, which is used to tell the story of the oxygenation of the ocean and atmosphere. A stromatolite boulder from the Atar formation, Mauritania, weighing 760 kg and displayed to expose its internal structure, reveals the nature of early life that provided this oxygen. A 2.4 meter-high section of glaciogenic conglomerate from the Huronian Supergroup, Ontario, records the Earth's first extensive and prolonged glaciation.

Why are there ocean basins, continents, and mountains? The center of this section focuses on deep Earth and plate tectonics. The critical samples are two boulders of red sandstone, which look ordinary, but convey complex ideas. One is the Old Red Sandstone from Scotland; the other, from New York, was deposited in the same basin about 360 million years ago and was separated from its

Scottish equivalent by the opening of the Atlantic. The four surrounding areas describe effusive volcanism, explosive volcanism, rock deformation and mountain building, and earthquakes. Among the outstanding samples is a massive, once-molten piece of sulfur from a fumarole on Kawah Ijen volcano, Java, portraying the nature of volcanic gases. Boulders of pumice, obsidian, welded ash, and basalt from Medicine Lake volcano illustrate the complexity of arc volcanism, and the story of Hawaiian volcanism is told through samples that include an entire lava tree.

In the area on earthquakes, a fault is exposed by its offset of a basalt dike in a large slab of granitic gneiss. Metamorphic rocks are offered as examples of the roots of mountains, and a boulder of garnet amphibolite from Gore Mountain, New York containing garnet megacrysts up to 50 cm across illustrates the gross changes in texture that can accompany metamorphism.

These bisected sections of sulfide chimneys are from that Mothra hydrothermal vent field on the Juan de Fuca Ridge.

Reading the rocks. This section features the story of the Grand Canyon. It also includes a description of Sm-Nd age dating, supported by a sample and by mineral separates used to construct an internal isochron of a 2.7 billion year old gabbro from the Stillwater complex, Montana.

What causes climate and climate change? This section explains how the ocean and atmosphere interact and circulate and how scientists study past climate, or paleoclimate, to understand how climate changes. The description of paleoclimate focuses primarily on a set of three one-meter segments of the GISP-2 Greenland ice core. (At present, these contain plastic ice: The display freezer is not yet functional.) Other samples of the climate record include lobate coral, tree rings, and lake and deep-sea sediment cores. One sample, a 225-kg ventifact from one of the dry valleys of Antarctica, is simply iconic, representing erosion and weathering.

Why is the Earth habitable? The most important samples in the hall are the centerpiece of this section: the dramatic "black smokers," or hydrothermal sulfide chimneys, collected from the Juan de Fuca Ridge by an American Museum of Natural History-University of Washington joint expedition. The first to be collected while actively venting hot fluid, these samples enable scientists to investigate the microbial community living in the rocks and its relations to the ambient thermal, mineralogical, and chemical conditions.

The chimneys are juxtaposed against a series of massive sulfide ores from the Kidd Creek Mine, Timmons, Ontario, which formed 2.7 billion years ago. The section also includes other ores from sedimentary, hydrothermal and magmatic environments. One of the most spectacular examples of a magmatic ore is a 1590 kg boulder from the Bushveld complex, showing the complete section of the Merensky reef, which provides about seventy percent of the world's platinum.

A cast recreation of the Kings Canyon fold from the Sequoia National Park in California.

Models and Reproductions

The hall contains a series of eleven full-scale reproductions of entire outcrops. The outcrops were painted with several layers of latex rubber, which was peeled off and used as a mold to fashion a life-like resin cast; the surface of the cast was then painted to match the outcrop. In the exhibit, casts are recognizable because their sides are exposed. The classic outcrops represented include a cliff face in Kings Canyon, California, showing chevron folds in metamorphosed siltstones and marbles in a roof pendant of the Sierra Nevada batholith; a section of the entire pyroclastic sequence of the 79 AD eruption of Vesuvius; the Hutton unconformity where it was first recognized at Siccar Point, Scotland; and the Glarus thrust, the base of an alpine nappe made notable by the nineteenth century Swiss geologist Arnold Escher von der Linth.

The hall also contains models, which are artistic, but nevertheless accurate representations of scientific processes and data. Among the most popular is a 107-centimeter diameter cast bronze globe of the Earth, carved from satellite data. The globe is touchable, inviting exploration by visitors. A reproduction of a carved plaster model of the Säntis Peak of eastern Switzerland is another favorite. The Swiss geologist Albert Hiem made the original model in 1899, and its large folds can be seen clearly by the distribution of rocks on the surface. A scientific sand model illustrates the gross structure of mountain belts created by continent-continent collision and also introduces the concept of scaling. A video associated with this model shows how it was made and includes time-lapse photographs illustrating the closing of an ocean basin, collision of continents, and growth of a mountain belt.

Videos and Sound

Videos are an essential part of the exhibition, but are used in supporting roles. A set of

A view of the north half of the Hall that highlights the range of exhibits including specimens, casts, models, and video.

seven "scientist-at-work" videos features scientists in the field pursuing activities ranging from monitoring the activity of Kilauea volcano, Hawaii, to studying the paleoseismicity of the San Andreas fault system, to recovering and analyzing the Greenland ice core. A second set of five videos, made in collaboration with Los Alamos National Laboratory, describes dynamic parts of the Earth and illustrates how numerical computer modeling and visualization are used in Earth science. The set includes stories on magnetic field generation, mantle convection, atmosphere circulation, ocean circulation, and flow of ground water. A third set of three interactive videos were created to investigate the carbon cycle, illustrate the drift of continents through time, and explore geo-chemical data from the ice core.

In addition, a 10-minute video entitled "Views of the Planet" is prominently projected on one wall. This video consists of various dramatic images of the Earth, each lasting for several seconds and dissolving into the next. It is accompanied by sound that includes low frequency rumblings played through benches. Other soundscapes help to set the tone elsewhere in the hall. In the middle of the hall, around the hemisphere, more gentle background sounds contribute to a contemplative mood; and in the area of the black smokers, a third soundscape evokes the mysterious quality of the deep ocean. Paradoxically, the soundscapes also serve to make the spaces quiet by disguising the ambient noise.

Conclusion

The Hall of Planet Earth has enormous potential to influence how the public thinks about the Earth and perceives the science. The exhibit is rooted in academic science and its functional, esthetic design helps visitors understand and experience that science in new and exciting ways. Visit the Hall of Planet Earth at www.amnh.org/rose/hope.

Resources

Suggested Reading

Bolt, Bruce A.
Earthquakes.
3rd Edition. New York: W.H. Freeman and Company, 1993.

Dr. Bolt is a renowned earthquake scientist. He was Director of the University of California Seismographic Stations for twenty-eight years. Bolt has made many post-earthquake investigations and has written and edited many textbooks on earthquakes, geology, and computers.

Dunning, F.W., P.J. Adams, J.C. Thackray, S. van Rose, I.F. Mercer and R.H. Roberts.
The Story of the Earth.
London: British Museum of Natural History, 1991.

This highly-recommended series of publications present Earth's history and dynamic systems for a general readership. It has a clear style and concise format without over-simplification.

Kingston, J. Longman.
Illustrated Dictionary of Geography.
Beirut: Longman Group UK Limited, Burnt Mill, Harlow, Essex and York Press, 1988.

A good basic reference book for geologic and geographic terms with numerous color illustrations.

Levin, Harold L.
The Earth Through Time.
5th Edition. Fort Worth, Texas: Saunders College Publishing, 1996.

This illustration-filled textbook about historical geology is extremely comprehensive and understandable.

Levy, Matthys and Mario Salvadori.
Why the Earth Quakes: The Story of Earthquakes and Volcanoes.
New York: W.W. Norton, 1995.

Amazing Earth events are depicted in this book featuring nice artwork.

Moores, E.M. and F.M. Wahl.
The Art of Geology.
Boulder, Colorado: The Geological Society of America, Inc., 1988.

Some of the most beautiful spots on the planet are captured in this photo album. Each landform is accompanied by short passages of text regarding the geology that created it.

Officer, Charles and Jake Page
Tales of the Earth: Paroxysms and Perturbations of the Blue Planet.
New York: Oxford University Press, 1993.

Some of the greatest events of environmental history, natural catastrophes, and disasters are described in this book.

Skinner, Brian and Stephen C. Porter
The Dynamic Planet: An Introduction to Physical Geology.
New York: John Wiley and Sons, Inc. 2000

This excellent college level textbook is linked to the Museum's Gottesman Hall of Planet Earth through a co-developed Web site (www.wiley.com/college/skinner). Fully illustrated, the book explores the processes that make Earth dynamic.

Videos

Volcanoes of the Deep
NOVA Video
To order, call 1-800-949-8670

See how black smoker sulfide chimneys can support oases of astounding creatures as well as hold clues to how life itself may have begun.

Warnings from the Ice
NOVA Video
To order, call 1-800-949-8670

NOVA follows scientists to the Antarctic as they battle extreme weather conditions in the hope of gathering data that will reveal new insight into the nature of global climate change.

Web Sites

American Museum of Natural History: Gottesman Hall of Planet Earth
www.amnh.org/rose/hope/

Explore the Museum's exciting new exhibition virtually. Follow the journey of the Hall from its conception, to the collection of rocks samples from around the world, to production and installation. You can explore current stories from the Earth Bulletin (http://earthbulletin. amnh.org, which presents scientific interpretations of current environmental and atmospheric events on our planet.

Atlas of Igneous and Metamorphic Rocks, Minerals, & Textures
www.geolab.unc.edu/Petunia/IgMetAtlas/mainm enu.html

This atlas provides views of rocks and minerals through a high-powered microscope. The site is hosted by the University of North Carolina.

Earth Science Resources on the Internet
www.lib.unc.edu/geolib/

This site, from the University of North Carolina at Chapel Hill, provides links to earth science sites, including seismology, weather and the environment, paleontology, and earth science schools and institutes.

Geologic Time
www.ucmp.berkeley.edu/help/ timeform.html

Explore a geologic timeline with links to a detailed description of each time period. The site, produced by the University of California at Berkeley, presents stratigraphy, ancient life, and plate tectonics for each time period.

A Geologist's Lifetime Field List
www.uc.edu/geology/geologylist/

A list of many of the most interesting geological wonders of the world, with links to most of them, can be found on this Web site, which was developed by the University of Cincinnati Geology Department.

Geologylink
www.geologylink.com

This site from the Houghton Mifflin Company includes the following sections: Earth Today, Virtual Field Trips, and an earth science Glossary. A Virtual Classroom area provides links to geology classrooms worldwide.

Geoscience Information Center
http://geosciences.org

Up-to-date articles in many fields of geoscience and information about current discoveries in weather, climate, paleontology, earthquakes, plate tectonics, hydrogeology, and more can be found on this site sponsored by the Scripps Institution of Oceanography at the University of California, San Diego.

Illustrated Glossary of Geologic Terms
www.geology.iastate.edu/new_100/glossary.html

Iowa State University developed this valuable and basic resource, a must for amateur geologists.

The Incorporated Research Institutions for Seismology (IRIS)
www.iris.edu

IRIS is a university research consortium dedicated to exploring the Earth's interior through the collection and distribution of seismographic data. IRIS programs contribute to scholarly research, education, and earthquake hazard mitigation.

Lamont-Doherty Earth Observatory (LDEO)
http://www.ldeo.columbia.edu/

LDEO is a research division of Columbia University whose scientists and graduate students work in seismology, marine and terrestrial geology, climate studies, atmospheric science, and oceanography, among others.

NASA: Earth from Space
http://earth.jsc.nasa.gov/

Access beautiful satellite images of Earth with this Web site tool.

NASA's Earth Observatory
http://earthobservatory.nasa.gov

On this Web site, NASA scientists explore the causes and effects of climatic and environmental changes. It also uses real satellite data to examine topics such as El Niño, climate modeling, and ozone in the stratosphere.

NASA: Teaching Earth Science
www.earth.nasa.gov/education
Produced as part of NASA's Education Outreach efforts, this site catalogs programs and resources for elementary through university levels.

National Oceanic and Atmospheric Administration (NOAA)
www.noaa.gov

NOAA warns the public of dangerous weather, charts our seas and skies, guides our use and protection of ocean and coastal resources, and conducts research to improve our understanding and stewardship of the environment.

National Parks Geology
www.aqd.nps.gov/grd/tour

The National Park Service developed these virtual Web tours of some of the most geologically interesting sites in the U.S. National Park system.

Scientific American
www.scientificamerican.com

Scientific American's Web site features scientific news. Articles on earth science appear frequently.

United States Geological Survey (USGS)
www.usgs.gov

The USGS serves the Nation by providing reliable scientific information to describe and understand Earth's geological and biological systems.

About the American Museum of Natural History

Founded in 1869, the American Museum of Natural History is one of the world's preeminent institutions for scientific research and education.

Today, under the direction of President Ellen V. Futter, the Museum's scientific, education, and exhibition staff are working to discover, interpret, and disseminate knowledge about human cultures, the natural world, and the universe. Prepared for the challenges of twenty-first century society, the Museum is committed not only to making contributions to science, but to improving science education and enhancing science literacy nationwide.

In 1997, the Museum launched the National Center for Science Literacy, Education, and Technology to take the Museum beyond its walls to a national audience. The National Center uses media and technology to connect people of all ages to real scientists and their work. The purpose of the National Center is to take the Museum's vast resources—collections of some 32 million specimens and artifacts, 43 exhibition halls, and a staff of more than 200 scientists, and over 130 years of expertise in educational programming—directly to classrooms, libraries, community centers, and homes throughout the country.

Contributors

Brian F. Atwater is a geologist with the United States Geological Survey and is an Affiliate professor in the Department of Geological Seattle. He uses geology to learn about the history of earthquakes and tsunamis in the Pacific Northwest, Alaska, Chile, and Japan. He became interested in geology while leading backpacking trips in New England as a high school senior. Atwater's co-authors, Tom Yelin and Craig Weaver are seismologists in Seattle and Jim Hendley is an editor in Menlo Park, California.

George Billingsley, raised in Arizona, was educated at Northern Arizona University, Flagstaff, Arizona and received both his B.S. and M.S. degrees in geology. Early in his career he became a professional Colorado River guide conducting geologic river trips through the Grand Canyon. He has worked as a research geologist with the Museum of Northern Arizona for a few years conducting geologic studies and geologic mapping of Grand Canyon National Park, Arizona, Canyonlands National Park, Utah, and Capitol Reef National Park, Utah. Since 1980, he has been a geologist by the U.S. Geological Survey where he continues geologic investigations in stratigraphy, geomorphology, and geologic mapping of the southern Colorado Plateau region. He has authored and co-authored 110 publications, 58 of which are geologic maps covering over 10,000 square miles of northern Arizona, and over 4,000 square miles of southern Utah.

Peter Bunge was born in Berlin, Germany. After receiving undergraduate degrees in geology and physics, he received his Ph.D. from the University of California at Berkeley in 1996. His thesis work was supported by the Los Alamos Branch of the Institute of Geophysics and Planetary Physics (IGPP) and the Advanced Computing Laboratory (ACL) of Los Alamos National Laboratory (LANL).

Bunge's interest in solving earth science problems with computers led to his creation of groundbreaking, high-resolution computational models. After spending two years as a post-doctoral fellow of the European Union at the Department of Numerical Modelisation at the Institut de Physique du Globe de Paris (IPGP), he joined Princeton University's Department of Geosciences in 1998 where he is an Assistant Professor of Geophysics. At Princeton his goal is to build a detailed Earth model that will enable him to understand the last 100,000,000 years of Earth's evolution and to make sense of the workings of the geodynamo.

Mark A. Cane is G. Unger Vetlesen Professor of Earth and Climate Sciences at Columbia University's Lamont-Doherty Earth Observatory. With Lamont colleague Stephen Zebiak, he devised the first numerical model able to simulate El Niño, and in 1985 this model was used to make the first physically-based forecasts of El Niño. Over the years the Zebiak-Cane model has been the primary tool used by many investigators to enhance understanding of ENSO. Cane has also worked extensively on the impact of El Niño on human activity, especially agriculture. His efforts over many years were instrumental in the creation of the International Research Institute for Seasonal to Interannual Climate Prediction.

Lowell Dingus was born in California, where he received his B.S. and M.S. in geology from the University of California at Riverside before receiving his Ph.D. in paleontology from the University of California at Berkeley. He served as scientific coordinator on the evolutionary exhibit entitled *Life Through Time* at the California Academy of Sciences before directing the renovation of the Fossil Halls at the American Museum of Natural History between 1987 and 1996. He has participated as chief geologist in the AMNH/Mongolian Academy of

Sciences expeditions, as well as the recent expeditions to Patagonia that discovered the first-known fossilized sauropod embryos and embryonic dinosaur skin. A Research Associate in the Department of Vertebrate Paleontology at the American Museum of Natural History, he also serves as president of the InfoQuest Foundation, which supports field research in geology and paleontology, as well as scientific and technological literacy among today's students. He has authored or co-authored numerous books on paleontology for both children and adults, including *What Color Is That Dinosaur?, Searching for Velociraptor, The Tiniest Giants, A Nest of Dinosaurs, Next of Kin, The Mistaken Extinction,* and *Discovering Dinosaurs,* which won a Scientific American Book for Young Readers Award.

Robert A. Fogel was born and raised in New York City. He received his Ph.D. in geology from Brown University in 1989. Fogel is currently a Research Scientist at the American Museum of Natural History where he studies the chemistry of the planets, meteorites, and stars. His interest in the Moon was cultivated early on when, as a young boy, he and his brother would glue themselves to their TV set to watch the Apollo missions to the Moon. Fogel has researched several aspects of lunar geology and is particularly interested in the chemistry and physics of lunar volcanic eruptions. He is currently funded by NASA to study the chemistry of the solar nebula through laboratory experiments and geochemical studies of 4.5-billion-year-old meteorites. "Bobby" is married and is the father of four young sons....who often look at the big white ball in the sky.

Gary A. Glatzmaier received a Bachelor of Science degree, magna cum laude, from Marquette University in 1971 and then served four years as an officer in the U.S. Navy teaching nuclear reactor physics. In 1980 he

received a Ph.D. in physics from the University of Colorado. After two post-doctoral positions, he spent sixteen years at the Los Alamos National Laboratory developing three-dimensional time-dependent computer models to study the internal structure and dynamics of planets and stars. In his studies of the Earth, he developed separate models that simulate the global circulation and convection in the Earth's atmosphere, mantle, and core. He produced the first successful computer simulations of the geodynamo, the mechanism in the Earth's fluid outer core that maintains the geomagnetic field. In 1995, based on these simulations, he predicted the super-rotation of the Earth's solid inner core, which seismic analyses now support. He is a Fellow of the Los Alamos National Laboratory and of the American Geophysical Union, and in 1996 won the Institute of Electrical and Electronics Engineers (IEEE) Sydney Fernbach Award for his geodynamo simulations. In 1998 he became a Professor of Earth Sciences at the University of California, Santa Cruz.

Steven L. Goldstein is an Associate Professor in the Department of Earth and Environmental Sciences at Columbia University. He received both his bachelor's and Ph.D. degrees from Columbia, and his master's degree from Harvard. After finishing his Ph.D. he was a Research Scientist in the Geochemistry Division of the Max-Planck-Institut für Chemie in Mainz, Germany, and returned to Columbia in 1996. His research covers a range of subjects in geochemistry, using isotopes to study the Earth's mantle, continents, oceans, and climate through time.

Charles F. Keller received a Bachelor of Science degree from Penn State University and a Ph.D. in astronomy and astrophysics from Indiana University. He is the Center Director of the Los Alamos National Laboratory branch of the University of California's Institute of

Geophysics and Planetary Physics (IGPP). Keller was a member of the United States delegation to the "Japan-U.S. Seminar on the Global Environment" in January of 1992. Keller participated in CEPEX, Central Equatorial Pacific Experiment on Ocean-Atmospheric Heat Transfer in March of 1993. He has a long-standing interest in climate physics in both ocean and atmospheric studies and global climate change. He participated in a debate with Fred Singer of the Science and Environmental Policy Project in October of 1998 at Los Alamos National Laboratory. He is an Adjunct Professor at the Los Alamos branch of the University of New Mexico.

Deborah S. Kelley received her geology degrees from the University of Washington and Dalhousie University where she began her work on analyses of fluids in the oceanic crust. She has been an Assistant Professor in the Department of Oceanography at the University of Washington since 1995. Her research focuses on the transport of gases from rocks deep within the ocean crust to the seafloor, and on the interplay among geological and biological processes in submarine hydrothermal vent systems. She has participated in numerous oceangoing research cruises that include diving in the three-person submarine *Alvin*, use of the remotely-operated vehicles Jason and ROPOS, and multiple Ocean Drilling Program legs. Kelley was a chief scientist on the sulfide recovery project in collaboration with the University of Washington and the American Museum of Natural History, and helped in the recovery of 4,000-pound, active black smokers from the seafloor.

David B. Loope grew up in the Shenandoah Valley of Virginia and studied Biology at Duke University. After a few years working as a ranger in Canyonlands National Park in southern Utah, he decided he wanted to learn more about geology. He studied the eolian sandstones of Canyonlands for his Ph.D. at the

University of Wyoming and has worked on windblown sediments and sedimentary rocks ever since. He lives with his wife and two sons in Lincoln, Nebraska, where he has taught in the Department of Geosciences at University of Nebraska since 1981.

Edmond A. Mathez is a Curator in the Department of Earth and Planetary Sciences at the American Museum of Natural History. He studies how magmas crystallize and the behavior of volatile compounds in the deep Earth. He is an expert on the geochemistry of platinum and the geochemistry of carbon. Mathez served as chief Curator of the Gottesman Hall of Planet Earth. He lives with his family in Nyack, New York.

Paul A. Mayewski was born in Edinburgh, Scotland and received a Bachelor of Science degree from the State University of New York at Buffalo, earned his doctoral degree from the Institute for Polar Studies at Ohio State University and has an honorary Ph.D. from Stockholm University in Sweden. He is the founder and Director of the Climate Change Research Center at the University of New Hampshire and more recently a Professor in the Institute of Quarternary Studies and Department of Geological Studies at the University of Maine. He is a Fellow and Citation of Merit Awardee of the Explorers Club and a Fellow of the American Geophysical Union. He has led more than thirty-five expeditions to Antarctica, the Arctic, and the Himalayas-Tibetan Plateau, and has published approximately 200 papers in professional journals. As Chief Scientist for the Greenland Ice Sheet Project Two (GISP2) he organized twenty-five universities in the pursuit of a climate record that has revolutionized our understanding of natural climate change. He organized fifteen nations in pursuit of records documenting the last few hundred years of climate change and change in the chemistry of the atmosphere, under a program that involves a series of oversnow traverses of Antarctica, called the International Trans Antarctic Expedition (ITASE). Currently, he is leading the U.S. field component and expeditions to the Himalayas.

Stephen J. Mojzsis, an Assistant Professor at the University of Colorado at Bolder, Department of Geological Sciences, is a specialist on the geology and geochemistry of the earliest rock record, and on the origin and evolution of life. Mojzsis has extensive field experience in site-mapping and sample-collecting in the oldest terrains known on Earth, particularly in the coastal regions of southern West Greenland, Western Australia, India, Southern Africa, and Northern America. He directs field and laboratory research at the University of Colorado, Boulder on the chemistry of the early Earth, of Mars, and of meteorites. When not in the lab, writing and presenting scientific papers, or on expeditions, he enjoys surfing the beautiful California coast and spending time hiking throughout the southwestern U.S. with his family.

Peter H. Molnar grew up in the east, and saw the Rocky Mountains for the first time at age nine; suddenly his life was focused without his knowing it. He studied Physics at Oberlin College, graduating in 1965, and then seismology in the Department of Geology at Columbia, where he obtained his Ph.D. He counts himself lucky to have been a student while plate tectonics was first recognized. He saw its early successes and failures, namely in continental regions where rigid plates are either too small and numerous to reveal simplicity or bounded by such wide, deforming margins to be insignificant. Following a post-doc at the Institute of Geophysics and Planetary Physics at Scripps Institute of Oceanography, he moved to MIT where he devoted most of his time to how continents deform, especially in Asia, where the mountains are highest and the plateaus widest. He commonly spends two to four months a year in the mountains carrying out geological or geophysical studies, and the rest of the year groping for ways to make sense of what he has seen, both in the field and with other data obtained at the armchair. Molnar is now a Professor of Geology in the Department of Geological Sciences of the University of Colorado, Boulder.

Rachel Oxburgh was born in Oxford, England and received her B.A. in Chemistry from Oxford University in 1988. She was granted a M.Sc. in Geology from the University of Wyoming in 1990 and a Ph.D. in Geochemistry from Columbia University in 1996. Oxburgh's research interests include the chemistry of natural waters and understanding the links between ocean chemistry and climate change. Now Lecturer in Geochemistry at Edinburgh University and Adjunct Associate Research Scientist at Columbia University's Lamont-Doherty Earth Observatory in New York, in her spare time Oxburgh enjoys running, climbing, and dancing.

Roberta L. Rudnick was born and raised in Portland, Oregon, where she studied geology at Portland State University. There, she witnessed the devastating 1980 eruption of Mt. St. Helens (from a safe distance) and was hooked on geology for life. She was awarded an N.S.F. graduate fellowship during her master's work at Sul Ross State University in Alpine, Texas, which had a major impact on her career path. She went on to earn a Ph.D. in geochemistry at the Australian National University where she began studying the composition of the lower continental crust. After two years as an Alexander von Humboldt fellow at the Max-Planck-Institut für Chemie in Mainz, West Germany, and five more years as a Research Fellow at the Australian National University, she took up a faculty appointment at Harvard University, where she was tenured as a full professor in 2000. In the same year, she moved to take up a professorship in the Department of Geology at the University of

Maryland at College Park. She has published extensively on the composition and evolution of the continents and their mantle roots. She lives with her husband, William F. Donough (also a geochemist) and her seven-year-old son, Patrick (a budding geochemist).

Haraldur Sigurdsson was born in Iceland, where he first encountered active volcanoes and became fascinated by their workings. He earned a Geology degree of Bachelor of Science from Queen's University, Belfast, Ireland, followed by a Ph.D. degree in Geochemistry from the University of Durham, England. He then worked for four years at the University of the West Indies, monitoring and surveying the numerous active volcanoes of the Caribbean region. He was appointed Professor at the University of Rhode Island in 1974, where he has pursued studies of both submarine and land volcanism worldwide, including research of active volcanoes in Africa, Indonesia, Italy, Central and South America, the South Pacific, and his native Iceland.

Robert D. van der Hilst was born in Oostzaan, The Netherlands, and was educated at Utrecht University, where he received Master of Science degrees in Geology and Geophysics, both cum laude, and his Ph.D. He held research positions at the University of Leeds, U.K., and at the Australian National University, Australia, where he spearheaded a nationwide program for seismic imaging. In 1996, van der Hilst joined the faculty of the Massachusetts Institute of Technology where he is now Professor of Geophysics. His major research focus has been the use of seismic imaging techniques to improve the understanding of structure and dynamic processes in Earth's deep interior. His multidisciplinary approach to solving outstanding problems has earned him several awards, including the Macelwane Medal from the American Geophysical Union in recognition of

outstanding young scientists and a Packard Fellowship from the David and Lucile Packard Foundation.

Martin H. Visbeck was born in northern Germany and received a Ph.D. in physical oceanography, magna cum laude, from Kiel University, Germany. In 1993 he was awarded a NOAA Global & Climate Change post-doctoral fellowship to work at the Massachusetts Institute of Technology. In 1995 he continued his career at Columbia University's Lamont-Doherty Earth Observatory and was promoted to Associate Professor in the Department of Earth and Environmental Science. Visbeck is an expert in deep water formation and deep convection in the ocean. He has participated in several seagoing expeditions on research vessels and ice breakers to the Greenland, Labrador, and Weddell Seas. Visbeck has also worked on several other topics related to the role of the ocean in the global climate. He has contributed to the latest assessments of the Intergovernmental Panel on Climate Change (IPCC) and plays an active role in shaping U.S. efforts in a recently launched international enterprise to study Climate Variability and Predictability (CLIVAR).

James Webster is the Chairman of the Division of Physical Sciences and a Curator in the Department of Earth and Planetary Sciences at the American Museum of Natural History in New York. Webster is the mineral deposits curator at the Museum, and his research interests involve the behavior of hot water and other gases in forming metal deposits and in driving volcanic eruptions. He has been with the Museum for nearly ten years, after two-and-a-half years of post-doctoral research with the University of Edinburgh and with the United States Geological Survey. He studied geology and geochemistry in graduate school at Arizona State University and the Colorado School of Mines.

Credits

Cover

Chris McKay, photo courtesy NASA Photo/Dale Anderson.

Contents

Earth, photo courtesy NASA.

Foreword

American Museum of Natural History—circa 1900, American Museum of Natural History.

Preface

Grand Canyon, photo by Denis Finnin, American Museum of Natural History.

Scottish dike, photo by Craig Chesek, American Museum of Natural History.

Section One

Banded Iron detail, photo by Craig Chesek, American Museum of Natural History.

Detail of fossilized stromatolite, photo by Craig Chesek, American Museum of Natural History.

The Moon, photo courtesy NASA.

Sequence of Moon formation, computer model visualizations courtesy of Eiichiro Kokubo and Hitoshi Miura, University of Tokyo.

Mt. St. Helens–pre-eruption, courtesy of the United States Geological Survey.

La/Nb of Average Continental Crust figure, courtesy Roberta L. Rudnick

Mt. St. Helens—post-eruption, courtesy of the United States Geological Survey.

Stromatolite image: © Fred Bavendam/Peter Arnold Inc.

Figures of changing concentrations of atmospheric carbon dioxide, courtesy of Stephen J. Mojzsis.

Mauritania stromatolite expedition images, photos by Craig Chesek, the American Museum of Natural History.

Scanning electron microscope image of zircon crystal, courtesy of Darrell Henry.

Ion probe lab at Wood's Hole Oceanographic Institute, photo courtesy Wood's Hole Oceanographic Institute.

Darrell Henry overlooking Quad Creek, photo courtesy Barbara Dutrau.

Paul Mueller with student Stephanie Ingle, photo courtesy Paul Mueller.

Sir Arthur Holmes, courtesy of University of Edinburgh, Department of Geology and Geophysics.

Section Two

Mantle xenolith, photo by Craig Chesek, American Museum of Natural History.

Kimberlite and diamond, photo by Denis Finnin, American Museum of Natural History.

Indonesian seismic station, photo by Denis Finnin, American Museum of Natural History.

Figure of Earth's internal structure, American Museum of Natural History.

Map of the migration of subduction zone locations, courtesy of Robert D. van der Hilst (from model of Lithgow-Bartelloni and Richards, 1977). Map adapted by David Zilkowski.

Figure of high wave speed structure, courtesy of Robert D. van der Hilst. Figure adapted by David Zilkowski.

Figures of 3D magnetic field structure, courtesy of Gary A. Glatzmaier. Figures adapted by David Zilkowski.

Figure of the churning Earth, American Museum of Natural History.

Images of models of mantle convection, courtesy of Peter Bunge.

Elise Knittle, photo by Don Fukuda, UC Santa Cruz.

Illustration of seismic waves, American Museum of Natural History.

Inge Lehmann, photo courtesy of B.A. Bolt.

Illustration of Earth's layers, American Museum of Natural History.

Section Three

Cliff wall in King's Canyon, photo by Denis Finnin, American Museum of Natural History.

Swiss mountain peak, photo by Craig Chesek, American Museum of Natural History.

Himalaya Mountains, photo courtesy Mark Marten/NASA/Photo Researchers, Inc.

Figure of Himalayan mountain-building, after Peter H. Molnar, illustration by Mick Ellison, American Museum of Natural History.

Dead forest in Washington, photo courtesy of Brian F. Atwater.

Figure of earthquake hazard zones for Pacific Northwest, courtesy of Brian F. Atwater. Figure adapted by David Zilkowski.

Figure of outlay of funds to improve infrastructure in the Pacific Northwest, courtesy of Brian F. Atwater. Figure adapted by David Zilkowski.

Close-up of lava, photo by Jackie Beckett, American Museum of Natural History.

Figure of age/distance relationship of Hawaiian Islands, courtesy of Steven L. Goldstein. Figure adapted by David Zilkowski.

Figure of subducting oceanic crust courtesy of Steven L. Goldstein. Figure adapted by David Zilkowski.

Mt. St. Helens erupting, photo courtesy United States Geologic Survey.

Figure of volcano explosiveness, American Museum of Natural History.

Figure of global distribution of explosive volcanoes, American Museum of Natural History.

Aerial image of the San Andreas transform fault, courtesy of the United States Geological Survey.

Thomas Rockwell in Mongolia, photo courtesy of Thomas Rockwell.

Sand model/Jacques Malavieille, photos by Jackie Beckett, American Museum of Natural History.

Harry Hess, photo courtesy of Princeton University, Department of Geosciences.

Section Four

Sandstone from Elliot Lake, photo by Craig Chesek, American Museum of Natural History.

Oviraptorid dinosaur fossil, photo by Denis Finnin, American Museum of Natural History.

Drawing of Grand Canyon by J.W. Powell from his book, Canyons of the Colorado, 1895.

Grand Canyon, photos by George Billingsley, courtesy United States Geologic Survey.

Ankylosaur fossil skeleton specimen, photo courtesy of Lowell Dingus.

Lowell Dingus, photo courtesy of David B. Loope.

David B. Loope, photo courtesy of Lowell Dingus.

Ankylosaur illustration by Ed Heck, American Museum of Natural History.

Mt. Rainier, photo by Jackie Beckett, American Museum of Natural History.

Tom Sisson on Mount Rainier, photo by Peter Haley / The News Tribune.

Dacite columns at Mt. Rainier, photo by Jackie Beckett, American Museum of Natural History.

Siccar Point, Scotland, photo by Craig Chesek, American Museum of Natural History.

James Hutton, portrait by Sir Henry Raeburn, courtesy of the Scottish National Portrait Gallery.

Section Five

Labate coral, photo by Craig Chesek, American Museum of Natural History.

Ice core extrusion photo by Anthony Gow, United States Army Corps of Engineers, Cold Regions Research and Engineering Laboratory.

Ka'imimoana at sea, photo by Linda Stratton, National Oceanographic and Atmospheric Administration.

Figure of ocean surface currents, American Museum of Natural History. Figure adapted by David Zilkowski.

Figure of the ocean circulation belt, American Museum of Natural History. Figure adapted by David Zilkowski.

TOA climate buoy, photo by Linda Stratton, National Oceanographic and Atmospheric Administration.

Figure of normal conditions, American Museum of Natural History. Figure adapted by David Zilkowski.

Figure of El Niño event, American Museum of Natural History. Figure adapted by David Zilkowski.

Figure predicting El Niño, courtesy Mark A. Cane, Lamont-Doherty Earth Observatory, Columbia University. Figure adapted by David Zilkowski.

Industrial smoke stacks, photo courtesy of Y. Hamel/Photo Researchers, Inc.

Figure of atmospheric concentration of carbon dioxide, courtesy of Charles F. Keller. Figure adapted by David Zilkowski.

Figure of oxygen isotope levels, courtesy of Charles F. Keller. Figure adapted by David Zilkowski.

Figure of measured global surface temperatures, courtesy of Charles F. Keller. Figure adapted by David Zilkowski.

GISP2 drilling dome, photo courtesy of Mark Twickler/University of New Hampshire.

Location map for deep drilling sites, courtesy of Paul A. Mayewski. Map adapted by David Zilkowski.

Figure of atmospheric composition, courtesy of Paul A. Mayewski. Figure adapted by David Zilkowski.

Figure of anthropogenic impact on atmosphere courtesy Paul A. Mayewski. Figure adapted by David Zilkowski.

Figure of varying levels of sulfate and nitrate with time, courtesy of Paul A. Mayewski. Figure adapted by David Zilkowski.

Tree ring section photo, courtesy Gordon Jacoby.

Gordon Jacoby, photos courtesy of Gordon Jacoby.

Perito Moreno Glacier, photo by Alberto Patrian, American Museum of Natural History.

Milutin Milankovitch, portrait by Paja Jovanovic, courtesy of John Imbrie, Brown University.

Rock outcrop, photo by Tina Gaud, American Museum of Natural History.

Section Six

Fragmented silver vein, photo by Craig Chesek, American Museum of Natural History.

Venus, Earth, Mars, photos courtesy NASA.

Figure of carbon cycle, by Mick Ellison, American Museum of Natural History.

Figure of radiations, by Mick Ellison, American Museum of Natural History.

Illustration of hydrothermal vent Godzilla by Veronique Robigou, courtesy Veronique Robigou, University of Washington.

Figure of global distribution of black smokers and cross section of hydrothermal fields, courtesy of Deborah S. Kelley. Figure adapted by David Zilkowski.

The Mothra Hydrothermal Fields, courtesy of Deborah S. Kelley.

Conditions to support life forms on sulfide chimneys and cross-section of sulfide chimney figures, courtesy of Deborah S. Kelley. Figures adapted by David Zilkowski.

The Climax Mine, photo by Denis Finnin, American Museum of Natural History.

James Webster, photo by Denis Finnin, American Museum of Natural History.

Polished slab cut from boulder, photo by Denis Finnin, American Museum of Natural History.

Microscopic inclusion in a crystal of quartz in a granite sample, photo by James Webster, American Museum of Natural History.

Mars with polar ice caps, photo courtesy United States Geological Survey.

Chris McKay, photo courtesy of NASA Photo/Dale Andersen.

Stromatolite photo by Craig Chesek, American Museum of Natural History.

Alvin at launch, photo by Rosamond Kinzler.

Veronique Robigou, photo by BLee Williams, Alvin pilot, Woods Hole Oceanographic Institution.

Stanley Miller in lab, photo courtesy SPL/Photo Researchers Inc.

Harold Urey, photo by Kemi, courtesy of the Nobel Foundation.

About the Hall of Planet Earth

About the Hall of Planet Earth photos by Dennis Finnin, American Museum of Natural History.

Glossary

abyssal ocean
The deep ocean that lies in water depths of 4,000 meters or deeper.

Acasta gneiss
The oldest rock dated on Earth, found in the Northwest Territories, Canada.

accretion
The process by which solid bodies gather together to form a continent or planet.

aerobic
Growing or thriving only in the presence of oxygen.

aerosols
Fine particles or liquid droplets suspended in the atmosphere, some of which are byproducts of industrial pollution.

air mass
A large, widespread body of air that has the same properties of humidity, temperature, and density (with only slight variations) throughout.

albedo
The proportion of incoming solar radiation reflected back into space either by the Earth's surface or by particles in the atmosphere.

amino acid
An organic compound containing an amino ($-NH_2$) and a carboxyl ($-COOH$) that link together to form proteins.

ammonia
(NH_3) A colorless, pungent gas, used to manufacture fertilizers and a wide variety of nitrogen-containing inorganic and organic chemicals.

ammonium
The chemical ion (NH_4^+) derived from ammonia that does not appear in a free state, but forms salts and compounds analogous to those of the alkali metals.

anaerobic
The condition of the absence of free oxygen.

analog model
A physical model that reproduces aspects of a specific phenomenon.

andesite
A dark-colored, fine-grained volcanic rock that is intermediate in composition (with respect to silica, magnesium and iron) between rhyolite (high silica, low magnesium and iron) and basalt (low silica, high magnesium and iron).

ankylosaurus
A genus of squat, quadrupedal, armored dinosaurs of the Cretaceous Period.

anorthite
A calcium-oxide-rich plagioclase feldspar mineral, typically white, that occurs in igneous rocks.

anorthosite
An intrusive igneous rock that consists chiefly of the mineral plagioclase feldspar.

Archean
The interval in Earth's history between 3.8 billion and 2.5 billion years before present.

atmosphere
The envelope of gases surrounding a planet, e.g. Earth's atmosphere consists predominantly of eighty percent nitrogen and twenty percent oxygen.

atmospheric circulation
Movement within the atmosphere caused by differences in air pressure.

atom
The smallest unit of an element that can still retain all of the characteristics and properties of that element, an atom consists of a dense, central, positively-charged nucleus surrounded by a cloud of electrons.

atomic weight
The total number of protons and neutrons in an atom. Also, the mass of an atom relative to the mass of the carbon atom having six protons and six neutrons, which is taken as 12.

bar
A unit of pressure equal to the weight of the atmosphere at one hundred meters above mean sea level.

basalt
A dark-colored, fine-grained volcanic rock that contains more iron and magnesium and less silica than andesite (rhyolite, andesite and basalt span the compositional spectrum of common volcanic rocks).

bedrock
The solid rock that underlies soil, sand, gravel, or other loose material.

bicarbonate
A salt containing a cation (any positively-charged ion) and the radical HCO_3, e.g., $NaHCO_3$.

bioherm
A mound-like or circumscribed mass of rock built almost exclusively of sedentary marine organisms that is embedded in a rock of different character.

biomass
The mass of living material.

biosphere
All living organizsms on Earth, including in its atmostphere, its waters, and in the solid Earth.

black smoker
A chimney-like structure made primarily of sulfide minerals that forms around hydrothermal vents on the ocean floor.

blue-yellow light
The most intense, dominant light from the solar spectrum, it is absorbed by plants' chlorophyll to make carbohydrates.

body wave
A seismic wave that passes through the Earth, traveling outward from an earthquake focus (the point of first release of energy that causes an earthquake).

bottom water
The deepest (and usually coldest) layer of the oceans consisting of the water that is in contact with the ocean floor.

butte
An isolated desert hill—usually flat-topped and steep-sided, smaller than a mesa.

calcium carbonate
($CaCO_3$) A colorless or white crystalline compound occurring naturally as the minerals calcite or aragonite in rocks like chalk, limestone, and marble.

carbon-14 dating
A method of determining the approximate age in years of a carbon-bearing object by measuring the decay of radioactive carbon-14.

carbonate
Any mineral compound that contains the anion (negatively charged molecule) CO_3^{-2}.

carbonate sedimentation
A process by which carbonate sediment is deposited.

carbon cycle
The combined processes by which carbon as a component of various compounds cycles between its major reservoirs: the atmosphere, oceans, living organisms, and solid Earth. The processes include photosynthesis, decomposition, respiration, sedimentation, lithification, burial, uplift, erosion, and volcanism.

carbon dioxide
(CO_2) An odorless, colorless, incombustible gas that is 1.5 times as dense as air and is formed during respiration, combustion, and organic decomposition. It is one of the principle greenhouse gases in Earth's atmosphere (after water vapor).

carbon monoxide
(CO) A colorless, almost odorless, poisonous, and flammable gas. It is a pollutant formed by the incomplete combustion of carbon-containing fuel.

carbonic acid
(H_2CO_3) A weak acid resulting from the solution of carbon dioxide in rain or groundwater.

catastrophism
The concept that major features in the Earth's crust such as mountains, valleys and oceans, have been produced by a few great catastrophic events, such as the Great Flood.

chemical equilibrium
The state in which forward and reverse chemical reactions occur at equal rates so that the concentration of the reactants and products does not change with time.

chemical weathering
The process of breaking down rocks or minerals at or near the Earth's surface by chemical processes, including hydrolysis, hydration, ion exchange, and oxidation.

chlorofluorocarbon
Synthetic chemical compounds used in refrigeration, solvents, and styrofoam manufacture. These compounds break down in the upper atmosphere and release chlorine atoms which destroy ozone.

chlorophyll
A group of green, light-collecting pigments found in green plants, algae, and some bacteria that in the presence of sunlight convert CO_2 and H_2O into carbohydrates.

climate
Average weather conditions of a region, including temperature, precipitation, and winds.

climatology
The scientific study of climate.

cobble
Any rock fragment larger than a pebble and smaller than a boulder, especially one that has been naturally rounded.

continental crust
The part of the Earth's crust that comprises the continents and is typically ~45 km thick. Together with the oceanic crust it makes up the outermost shell of the solid Earth.

continental drift
The slow, lateral movements of continents across the surface of the Earth.

convection
The process by which hot, less dense material rises upward and is replaced by cold, more dense, downward-flowing material.

convection cells
The currents that are set up by convection.

core
The spherical mass, largely of metallic iron and nickel, at the center of the Earth. The outer core extends from 2,900 kilometers to 5,100 kilometers from Earth's surface and is molten. The inner core, from 5,100 kilometers to the center of Earth at 6,400 kilometers, is solid.

core-mantle boundary
The separation between the liquid metal of the outer core and the solid rock of the lower mantle (also called the Gutenberg Discontinuity).

Coriolis force
An affect that causes any body that moves freely with respect to the solid Earth to veer to the right in the Northern Hemisphere and to the left in the Southern Hemisphere.

cosmic radiation
High-energy, subatomic particles from outer space, which bombard Earth's atmosphere. Most cosmic radiation is absorbed in Earth's upper atmosphere.

country rock
The older, preexisting rock that encloses or is traversed by an igneous intrusion or a mineral deposit.

crevasse
A deep, almost vertical, crack or split in the upper part of a glacier.

crossbeds
Layers that are inclined with respect to a thicker layer within which they occur.

crust
The outermost and thinnest of the solid Earth's layers, which consists of rocky material that is less dense than the rocks of the mantle below.

crustal thickening
The thickening of the continental crust when two continents collide to create mountains.

cryptoendoliths
Microscopic organisms that grow within rocks.

crystal
Any homogeneous solid with a regularly repeating atomic arrangement that may be expressed by plane faces; and a characteristic composition.

crystal fractionation
The separation of crystals of one composition from magma of another composition, thus causing the residual magma to change composisiton.

D" (D double prime)
The name for the lowermost 100–200 kilometers of the mantle where physical and chemical properties change rapidly.

debris flow
The down-slope movement of a mass of unconsolidated rock, sand, and dirt, more than half of which is coarser than sand.

decadal
On the time scale of decades, or tens of years.

deep water
The layer of the ocean between the intermediate water and the bottom water.

deglaciation
The uncovering of a landmass by retreat or melting of glacial ice.

dendroclimatology
Measurement of time intervals by counting the annual rings of trees; applicable to the last 3,000–4,000 years.

denudation
The progressive lowering of the Earth's surface by erosion, weathering, mass wasting, and transportation.

deposition
The act of laying down rock-forming matter (sediment) by a natural process, such as wind, water, or ice.

deuterium
An isotope of hydrogen with one proton and one neutron in the nucleus, giving it an atomic weight of 2.

differentiation
The process by which a magma changes composition, for example, by crystal fractionation. In reference to planets, the process of forming concentric layers of different composition, usually by sinking of dense material and floating of light material.

dipole
Two poles of opposite charge but equal magnitude.

discontinuity
A surface at which seismic wave velocities abruptly change due to changes in physical properties of the material the waves are traveling through (such as density or composition).

ductility
The ability of a material to irreversibly deform without rupture.

earthquake
A sudden motion or trembling of the Earth's crust caused by the passage of seismic waves radiated from a fault along which sudden movement has occurred.

electron
Negatively charged atomic particle.

electron microscope
A type of microscope that uses electrons rather than visible light to produce magnified images.

element
A substance composed of atoms having an identical number of protons in each nucleus. An element cannot be separated into simpler substances by chemical means. A trace element is one present in only small quantities, less than 0.1 percent by weight.

elevation
The height of an object above a particular reference level, usually sea level.

elliptical orbit
A planetary orbit that traces an elliptical course.

El Niño
A climactic event that generally occurs every four to twelve years, in which warming of the ocean surface off the western coast of South America disrupts the normal pattern of the upwelling of cold, nutrient-rich water. It causes fish and plankton to die and affects weather over much of the Pacific Ocean and around the world.

El Niño Southern Oscillation (ENSO)
A global set of climatic conditions made up of two components: an oceanographic one called El Niño, and an atmospheric one, the Southern Oscillation.

erosion
The complex group of related process by which rock is broken down physically and chemically and the products are moved. Agents of erosion include water, wind, and ice, as well as biological processes.

eukaryote
A cell that possesses a defined nucleus surrounded by a membrane. Protists, fungi, plants, and animals are eukaryotes.

exobiology
The branch of biology that deals with of the origin and distribution of possible life on other planetary bodies.

exsolution
The process by which one distinct phase of matter separates from another.

extraterrestrial
Occurring or originating away from Earth and its atmosphere.

fault
A rock fracture or fracture zone along which there has been movement.

feldspar
A group of rock-forming minerals that contain aluminum and silica and varying amounts of sodium, potassium, and calcium. Feldspar is the most abundant mineral group and makes up sixty percent of the Earth's crust. Feldspar is typically white, pink, or clear.

filamentous bacteria
Bacteria comprised of long threadlike microorganisms, often in interwoven colonies.

fission
The process, either spontaneous or induced, of the splitting of a nucleus into two or more large fragments of comparable mass plus some neutrons. It is accompanied by the release of vast amounts of energy.

folding
The bending of layers of rock, usually due to compression.

foraminifera
Microscopic, hard-shelled, marine to brackish (slightly salty water) single-celled organisms. Foraminifera create shells of calcium carbonate or make a shell by gluing tiny particles together; the composition of the shells are highly sensitive to temperature change.

fossil
Any remnant or trace of an organism of a past geologic or prehistoric age, such as a skeleton or the imprint of a leaf, embedded and preserved in the Earth's crust.

fossil fuels
A general term for any hydrocarbon deposit, such as oil, coal, or natural gas, derived from organic matter of a previous geologic time and used for fuel.

fossiliferous
Containing fossils.

geochemistry
The study of the distribution and amounts of chemical elements in the various systems that comprise the Earth.

geodynamics
The study of processes, such as the movement of material, in the Earth's interior.

geologic time
The period of time covering the physical formation and development of Earth, as recorded within the succession of rocks.

geology
The study of Earth, its history, its composition, its structure, and the dynamic processes that shape it.

geomagnetism
The study of the magnetic activity of Earth and its atmosphere.

geomorphology
The study of landforms and their origin on the surface of Earth and other planets.

geophysics
The study of physical properties of Earth.

GHGs
See **greenhouse gases**.

gigapascal
A billion pascals; a pascal is a unit of pressure (force per unit area).

glacial period
See **ice age**.

glaciation
The formation, advance, and retreat of glaciers through time. Glaciation of a region refers to the accumulation of ice over that region.

glacier
A large mass of ice, air, water, and rock debris formed at least partially on land which flows by internal deformation in response to gravity. Glaciers include small valley glaciers, ice streams, ice caps, and ice sheets.

granite
An intrusive igneous rock, usually light-colored. Granites commonly contain high amounts of quartz and feldspar. Micas, such as muscovite and biotite, may also be present. The extrusive igneous counterpart to a granite is a rhyolite.

greenhouse gases
Gases, primarily water vapor, carbon dioxide, and methane, that increase global temperatures by absorbing outgoing radiation emitted by the Earth's surface. Also called GHGs.

guyot
A submerged, flat-topped volcanic mountain formed in deep oceans.

gyre
The large, roughly circular current that is the main feature of wind-driven surface circulation found in most major ocean basins.

half-life
The time required for half of the initial number of atoms of a radioactive parent element (such as ^{14}carbon) to change into atoms of its daughter element (^{14} nitrogen).

headward erosion
The process by which the higher, originating end of a river wears away the rock around it, thereby lengthening tributary streams.

Holocene epoch
The most recent geologic epoch of the Quaternary Period extending from the end of the Pleistocene (11,000 years ago) to the present.

hotspot
A fixed point on the Earth's surface defined by long-lived volcanism.

hydrosphere
All of Earth's water, including the oceans, lakes, streams, water underground, and all the snow and ice, including glaciers.

hydrothermal fluids
Hot brines either given off by cooling magmas, or produced by reactions between hot rock and circulating water, that concentrate minerals in solutions.

ice age (glacial period, glacial epoch)
Recurring periods in Earth's history when the climate was colder and glaciers expanded to cover larger areas of the Earth's surface. The most recent ice age occurred during the Pleistocene epoch.

ice core samples
Samples of layered ice from glaciers which may contain dust, chemicals, and gases that have been deposited with snow over hundreds of thousands of years. These layers reveal past climate characteristics and many of their potential causes.

ice sheet
A large mass of ice thick enough to cover the topography under it. Ice sheets are large enough to deform and move with gravity.

igneous
Rock formed by solidification from a molten or partially molten state.

inclusion
Any solid, liquid, or gaseous foreign substance trapped inside a mineral or rock. Also refers to a fragment of an older rock embedded within an igneous rock.

infrared
Electromagnetic radiation lying in the infrared spectrum with wavelengths greater than red light (longer than the longest visible wavelengths).

inner core
The central or innermost part of the Earth's core, extending from a depth of 5,100 kilometers to the center of Earth at 6,400 kilometers. It is believed to be solid, as opposed to the outer core, which is liquid.

insolation
The solar radiation falling on Earth's surface or its atmosphere. A contraction of "*incoming solar radiation.*"

interdune valleys
Valleys or depressions between sand dunes.

interglaciation (interglacial period)
The time between glaciations when Earth's climate is warmer and the ice sheets have withdrawn, or retreated, from large areas of the continents. The present time is part of an interglacial period.

Intermediate water
Layer of ocean water above the deep water and immediately below the mixed layer of water at the ocean surface.

ion
A positively or negatively electronically-charged atom or molecule.

island arc
A curving group of volcanic islands parallel to a deep-sea trench. Island arcs and the deep-sea trench mark the location where oceanic crust is being subducted under oceanic crust. The Aleutian Islands are an island arc.

isostasy
The mechanism whereby areas of the crust rise or subside until the mass of their topography is buoyantly supported or compensated by the thickness of crust below, which "floats" on the denser mantle below.

isotope
One or more atoms of the same chemical element that differ in atomic weight because they have different numbers of neutrons. The atomic weight of the isotope is written in superscript to the left of the chemical symbol, such as ^{14}C.

Jacutophyton
A stromatolite with a central conical structure that has branches coming off all around it.

lahar
A fast-moving mudflow of unconsolidated volcanic ash, dust, breccia, and boulders mixed with rain, melting ice, or the water of a lake displaced by a lava flow.

landslide
Any perceptible downslope movement of a mass of bedrock or unconsolidated rock, sand and dirt, or a mixture between the two.

La Niña
The movement of colder-than-normal surface waters across the Pacific towards Indonesia that sometimes follows in the year after an El Niño event.

latitude
Imaginary lines that allow measurement of position north or south of the equator ("horizontal"). Latitude is measured in degrees (at the equator one degree = 60 nautical miles, or 111 kilometers). The equator is at a latitude of 0° and the poles lie at latitudes of 90° north (North Pole) or 90° south (South Pole).

lava
Molten rock that erupts onto the Earth's surface through a volcanic vent or fissure.

lava dome
A dome-shaped mass of sticky, gas-poor lava erupted from a volcanic vent, often following a major eruption.

Lehmann Discontinuity
The boundary between the Earth's solid inner core and liquid outer core. Named for the Danish seismologist Inge Lehmann.

limestone
A common sedimentary rock consisting mostly of calcium carbonate, predominantly formed from the skeletons of marine organisms, calcareous sand, microorganisms, shell fragments, and coral.

lithosphere
The solid, outermost shell of the Earth (~100 km thick), where rocks are more rigid than those below. The lithosphere is made up of the uppermost mantle and the crust.

lodestone
A piece of naturally occurring magnetic iron oxide (Fe_3O_4, the mineral magnetite).

longitude
Imaginary lines that wrap around Earth intersecting at the north and south geographic poles. Lines of longitude are numbered from 0° (Greenwich Meridian, passing through London, England) to 180°. Longitudes are designated east if they fall east of the Greenwich Meridian, and west if they fall west of the Greenwich Meridian.

lower mantle
The part of the mantle that lies below a depth of 1,000 kilometers. In this layer, seismic velocity increases slowly with depth.

Lunar Highlands
Light-colored regions of the Moon that are mostly composed of the rock anorthosite.

Lunar Magma Ocean
A sea of molten rock that existed shortly after the Moon formed which covered the entire Moon and extended down to depths of several hundred kilometers.

Lunar Mare
Dark-colored, low lying regions of the Moon comprised mostly of basalt.

magmatism
The development and movement of magma, and its solidification to form igneous rock.

magma
Molten rock within the Earth. Igneous rocks form when magma cools and crystallizes.

magnetic field
The region of influence of a magnetized body, such as Earth.

magnetic pole
Either of two variable points on Earth, close to but not coinciding with the geologic North and South Poles, where the Earth's magnetic field is most intense and towards which a compass needle points.

mantle
The layer within the interior of the Earth that lies between the crust and the core.

mass spectrometer
An instrument for separating atoms or molecules according to mass and counting them. Typically used for determining isotopic abundances.

mass wasting
The process in which a large amount of loosened soil and rock is transported downslope under the direct influence of gravity.

mesa
An isolated, steep-sided, flat-topped landmass rising above the surrounding geography.

Mesozoic
The Era of geologic time (from the end of the Paleozoic, 248 million years ago, to the beginning of the Cenozoic Era, 65 million years ago), including the Triassic Period, the Jurassic Period, and the Cretaceous Period. This Era is characterized by the development of flying reptiles, birds, and flowering plants, and the appearance and extinction of dinosaurs.

metamorphic rocks
Rocks that form as the result of transformation from other rocks. Metamorphic rocks are created when igneous, sedimentary, or metamorphic rocks change in response to extreme temperatures and/or pressures, but do not completely melt. Metamorphic rock types include slate, schist, and gneiss.

meteorite
A fragment of rock that has reached the Earth's surface from beyond the Earth's atmosphere.

meteorology
The study of Earth's atmosphere and the motion within the atmosphere. Meteorology includes understanding the aspects of the atmosphere for weather forecasting.

methane
An odorless, colorless, flammable gas (CH_4). Methane is an important source of hydrogen and a wide variety of organic compounds, as well as a principle constituent of natural gas.

mid-ocean ridge
A nearly continuous undersea mountain chain that marks the location where tectonic plates (pieces of the lithosphere) are diverging or moving apart. Mid-ocean ridges are the locations of creation of new ocean crust.

Milankovitch cycles
The three cycles related to variations in the Earth's rotational and orbital characteristics around the sun that are believed to influence the occurrence of ice ages, occurring at 100,000,

41,000, and 22,000 years. Named after Serbian mathematician Milutin Milankovitch.

mineral
A naturally-occurring, homogeneous inorganic element or compound having a definite chemical composition and orderly internal structure, crystal form, and characteristic chemical and physical properties.

mineralogy
The study of minerals, including their formation, occurrence, properties, composition, and classification.

mineral physics
The study of the physical properties of minerals.

Mohorovicic discontinuity (Moho)
The boundary between the crust and mantle marked by abrupt increases in seismic velocities. The Moho occurs at about five to ten kilometers beneath the ocean floor and about forty kilometers beneath the continents (although it may reach sixty kilometers or more under some mountain ranges).

molecule
The smallest unit of matter into which an element or a compound can be divided and still retain its chemical and physical properties. It consists of a single atom or group of like or different atoms bonded together by chemical forces.

molybdenite
A metallic, lead-gray, hexagonal mineral made of molybdenum and sulfur (MoS_2).

monsoon
A wind system that influences climatic regions and reverses direction seasonally. The Indian Monsoon brings wind from the southwest or south and brings great annual variation of rainfall to southern Asia and along the coasts of other regions of the Indian Ocean.

moraine
An accumulation of poorly sorted glacial sediments deposited beneath or at the margin of a glacier and having a surface form that is unrelated to the underlying bedrock.

neutron
An electrically neutral atomic particle that is stable when bound in an atomic nucleus. Neutrons are present in all known nuclei except the lightest isotope of hydrogen.

nitrous oxide
A colorless gas or liquid (N_2O), soluble in water or alcohol. It is a product of the combustion of fossil fuel.

Northern Hemisphere
The half of the planet north of the equator.

nuclear fusion
A nuclear reaction in which two light nuclei combine, at extremely high temperatures, to form a heavier nuclei and simultaneously release a vast amount of energy.

ocean basin
The area of the seafloor from the base of the continental margin (generally at the beginning of the continental rise) to the mid-ocean ridge.

oceanic crust
Part of the crust that forms the upper part of the rigid lithosphere, the outer layer of the Earth. Oceanic crust is formed at mid-ocean ridges and is denser and typically thinner than continental crust. Oceans are underlain by oceanic crust.

oceanic trench
A narrow, steep-sided, elongated depression of the deep-sea floor.

oceanography
The study of the ocean, including the physical properties of the ocean such as the currents and waves (physical oceanography), the chemistry of the ocean (chemical oceanography), the geology of the seafloor (marine geology), and the organisms that carve their niche within the ocean realm (marine biology and marine ecology).

outer core
The upper or outermost part of the Earth's core extending from 2,900 to 5,100 kilometers below the surface. It is believed to be liquid because it significantly reduces P-wave (compressed waves) velocities and does not transmit S-waves (shear waves).

Oviraptor
A primitive theropod dinosaur that developed from the carnivorous raptor theropods during the Late Cretaceous Period. The Oviraptor's skull was light in weight and its bones thin, much like modern birds. It had a short, thick, parrot-like beak and ran on its hind legs.

oxidation
A process in which a chemical element loses electrons.

ozone
A chemically active bluish gas that is made of molecules of three oxygen atoms (O_3). In the stratosphere, ozone acts as a protective barrier for Earth's surface by absorbing much of the potentially damaging ultraviolet radiation that comes from the Sun. In the troposphere, ozone acts as a harmful pollutant.

ozone layer
A layer in the stratosphere that contains about ninety percent of Earth's ozone. The ozone layer occurs approximately twenty-five kilometers (sixteen miles) above the surface of Earth.

Pacific Rim
The countries and landmasses surrounding the Pacific Ocean.

paleoclimate
The past climate.

paleontology
The study of the plants and animals of the past.

paleoseismology
The study of past earthquakes.

paleothermometer
The method or substance used to determine temperatures at a given time in the past.

patina
The colored film or thin layer on the surface of a rock produced by chemical weathering.

peridotite
A coarse-grained igneous or metamorphic rock composed primarily of olivine, with or without other minerals such as pyroxenes, amphiboles, or micas.

permafrost
Permanently frozen soil or subsoil, occurring throughout the polar regions and locally in perennially frigid areas. Its thickness ranges from thirty centimeters to over 1,000 meters and underlies approximately one-fifth of Earth's land area.

perovskite
The mineral $CaTiO_3$. At high pressure the mineral pyroxene ($MgSiO_3$) transforms to a form having the same structure, or arrangement of atoms, as perovskite. This $MgSiO_3$ perovskite makes up most of the lower mantle.

photic zone
The layer in a body of water or aquatic system that is penetrated by sunlight sufficient for photosynthesis to occur, extending as much as 150 meters below the surface.

photolysis
Chemical decomposition of a substance into simpler units as a result of its absorbing light or other radiant energy.

photosynthesis
The chemical process in which green plants (and blue-green algae) make carbohydrates from carbon dioxide and water using sunlight or light as an energy source. Most forms of photosynthesis release oxygen as a byproduct, the chief source of atmospheric oxygen.

physical weathering
The processes that mechanically break up rocks into fragments, such as the movement of water, wind and ice, and frost action. Physical weathering is also known as mechanical weathering.

physics
The study of matter, energy, motion, force, and their interrelationships.

planetesimal
Any of innumerable small bodies or satellites that are the precursors of a planet during the early stages of the solar system. A planetesimal can also be the fragmented result of a cataclysmic collision between a planet and another celestial body.

plasma
An electronically neutral, highly ionized gas composed of ions, electrons, and neutral particles. It is a state of matter distinct from solids, liquids, and normal gases that exists in extreme heat.

plate
One of several large, mobile pieces of the Earth's lithosphere adjoining other plates along zones of seismic activity.

plateau
A relatively elevated, comparatively level expanse of land with at least one abruptly steep side. It is higher than a plain and more expansive than a mesa.

plate spreading
The separation of two lithospheric plates, as occurs at the mid-ocean ridge.

plate tectonics
The theory and study of plate formation, movement, interaction and destruction; the attempt to explain seismicity, volcanism, mountain-building, and paleomagnetic evidence in terms of plate motions.

Pleistocene
The earlier of the two epochs of the Quaternary Period, ending around 11,000 years ago and characterized by the alternate appearance and recession of northern glaciation and the appearance of progenitors of human beings.

plumb bob
A pointed, tapering weight attached to a plumb line, used to measure the verticality of objects.

plume
A narrow, focused upwelling of unusually hot solid rock moving upward through the Earth's mantle.

Precambrian
The largest division (ninety percent) of geologic time, from 4.55 to 0.55 billion years ago. It precedes the Cambrian Period and is subdivided into the Hadean, Archean and Proterozoic Eons.

precipitation
The transfer of moisture from the atmosphere to the surface of Earth, usually as rain, snow, and ice.

primordial
Happening first in a sequence of time. Belonging to or remnant of the first stage of development, as in the formation of the Earth.

Proterozoic Era
The more recent of three divisions of the Precambrian, from 2.5 billion to 543 million years ago.

protons
A stable, positively-charged atomic particle found in the nuclei of matter.

protosun
A contracting gas cloud which was the earliest stage of the Sun's formation. This stage took place before the temperature and pressure of the interior became so high as to cause thermonuclear reactions.

pumice
Light-colored, porous, glassy fragments of lava typically with the composition of rhyolite.

P-wave
Primary or compressional seismic wave. It is the fastest of the seismic waves, traveling 5.5–7.2 kilometers per second

in the crust and 7.8–8.5 kilometers per second in the upper mantle.

pyrite
A common, brass yellowish-white, metallic, cubic mineral made of iron and sulfur (FeS_2). It is used to produce sulfur dioxide for sulfuric acid. Popularly known as "fool's gold," pyrite has metallic luster and has been mistaken for gold, which is more yellow, softer, and heavier than pyrite.

pyroclast
A molten fragment of pumice and ash ejected during a volcanic eruption.

pyroclastic flow
A high speed avalanche of hot volcanic ash, rock fragments, and gas.

quartz
A typically clear, hard mineral composed of silicon and oxygen. It is common in igneous, metamorphic, and sedimentary rocks, including sandstone and granite.

radioactivity
The spontaneous emission of energetic particles and/or radiation (including alpha particles, nucleons, electrons, and gamma rays) from either unstable atomic nuclei or as the consequence of a nuclear reaction.

radiocarbon dating
A dating method used to quantitatively measure the age of organic matter (such as bone, shell, or wood). The method can by applied to materials formed within the last 50,000 years or so.

radiometric dating
Determining the absolute age of rocks by measuring the decay of naturally occurring radioactive isotopes like carbon-14 or long-life isotopes of potassium, thorium, and uranium.

rhyolite
A fine-grained volcanic igneous rock usually light in color. Rhyolites commonly contain high amounts of silica, and low amounts of iron and magenesium. The fine-grained nature is a result of rapid cooling at the surface of Earth.

Richter scale
An exponential scale ranging from 1 to 9 that measures the amount of energy released during an earthquake.

Ring of Fire
The name of the extensive area of volcanic and seismic activity that roughly coincides with the borders of the Pacific Ocean.

rock
Any naturally formed aggregate of one or more minerals, such as granite, shale, or marble.

rockfall
The free falling of detached bodies of bedrock from a cliff or steep slope.

salinity
The amount of dissolved salts in water.

sand slurry
A thin mixture of sand and a liquid, usually water.

sandstone
A clastic, sedimentary rock formed by the consolidation and compaction of sand (primarily quartz) in a matrix of silt or clay and held together by a natural cementing material, such as silica, iron oxide, or calcium carbonate.

seafloor spreading
The movement of plates and the formation of new ocean crust at divergent plate boundaries, as at mid-ocean ridges.

seamount
An elevation of the seafloor, 1,000 meters or higher, having either a flat-topped or peaked summit below the surface of water, usually volcanic in origin.

sediment
Unconsolidated particles, ranging from clay-size to boulders produced by the breakdown of rocks that may be carried by natural agents (wind, water, and ice) and eventually deposited to form sedimentary deposits. Organisms and chemical precipitation can also produce sediment.

sedimentary rock
A rock formed by the consolidation or cementation of sediment particles, or chemically precipitated at the depositional site.

sedimentation
The process of sediment accumulation.

seismic tomography
An imaging technique using speeds of seismic waves to infer the three-dimensional internal density structure of the Earth.

seismic wave
Elastic vibration that travels through the Earth caused by an earthquake or a manmade explosion.

seismogram
The record of seismic waves on a seismograph after they have traveled through the Earth and arrived at a given seismic station. This record can be used to determine the location and strength of an earthquake.

seismology
The study of earthquakes and the mechanical properties of the Earth.

shale
A fine-grained, finely laminated, sedimentary rock, formed by the compaction of clay, silt, or mud. Its laminated structure makes it fissile, or easily split along close-spaced planes, especially on weathered surfaces.

sheetwash
Erosion of the ground surface by thin sheets of rainwater. Also known as "sheet erosion."

shock wave
A large amplitude compressional wave formed by an explosion or by supersonic motion, such as breaking of rock and movement along a fault within the Earth.

short-wavelength visible radiation
See **visible light**.

silica
The compound silicon dioxide (SiO_2). Silica is an important component of many rocks and minerals. It can be found in several forms, including quartz and opals.

silicate
Any mineral with a crystal structure containing silicon and oxygen (SiO_4) tetrahedra either isolated or joined through one or more oxygen atoms to form groups or three-dimensional structures with metallic elements.

silt
A sedimentary rock fragment or mineral particle that is finer than sand but coarser than clay.

siltstone
A fine-grained rock of consolidated silt with the texture and composition of shale, but lacking its fine lamination or fissility (ability to be easily split).

slump block
A large piece of rock that has broken off from the bedrock but has not yet shattered and broken. In the Grand Canyon, some slump blocks are as large as 2,000 meters long and 1,000 meters thick.

solar nebula
The cloud of interstellar gas and dust out of which the Sun and planets of the Earth's solar system were formed, roughly 4.5 billion years ago.

solar radiation
Energy from the Sun in the form of electromagnetic waves.

Southern Hemisphere
The half of the planet south of the equator.

spectroscope
An instrument that separates light by its wavelength. It is used to observe the light spectrum.

spiral galaxy
A system of roughly one hundred million stars arranged as a disk with a nucleus of older stars and spiral arms consisting mainly of dust, gas, and young stars.

spreading center
A linear region on the seafloor from which adjacent crustal plates are moving apart and along which magma rises to form new oceanic lithosphere.

statistical model
A model or simulation based on a sequence of past observations.

strata
Layers of sedimentary rocks which might contain differences in texture, color, fossil content, or material type.

stratification
Deposition of sediment in layers or strata.

stratosphere
The layer of the atmosphere above the troposphere. The stratosphere extends from approximately ten to fifty kilometers above Earth's surface.

stromatolite
A widely distributed sedimentary structure consisting of laminated carbonate or silicate rocks. It is produced over geologic time by the trapping, binding, or precipitating of sediment by groups of microorganisms, especially blue-green algae, in shallow, warm waters.

subcontinent
A large, relatively distinct landmass, such as India, which is part of a continent but geographically considered an independent entity.

subduction
The process of one tectonic plate moving beneath another tectonic plate at a convergent margin (where two plates collide). If continental lithosphere and oceanic lithosphere converge, the less dense continental lithosphere rides over the oceanic lithosphere and the denser oceanic lithosphere is subducted.

subduction zone
A long, narrow belt where subduction occurs, usually marked by island arcs such as the Aleutians, or volcanically active mountain chains such as the Andes.

sulfate
A negatively charged molecule containing sulfur and oxygen (SO_4^{-2}). Major sources of sulfates are fossil fuel burning and volcanic activity.

sulfide chimneys
A tower or chimney composed primarily of sulfide minerals and built up on the deep seafloor by mineral deposits precipitated from hydrothermal fluid ejected from a hydrothermal vent. The hydrothermal fluid contains hydrogen sulfide and high concentrations of metals such as iron, copper, and zinc, which can build a sulfide chimney around the vent as high as forty-five meters.

superposition (law of superposition)
A general law upon which all geologic chronology is based that states that in undisturbed, stratified sedimentary rocks (or of extrusive, igneous rocks) the lowest layers were deposited the earliest and are the oldest, and the top layers were deposited later and are therefore younger.

S-wave
A secondary or shear seismic wave. It does not travel through liquids and therefore cannot penetrate the outer core of the Earth. It travels 3.0–4.0 kilometers per second in the crust and 4.4–4.6 kilometers per second in the upper mantle.

talus
Accumulation of fragmented rock debris, usually coarse, angular, and of different sizes, found at the base of valley slopes and walls and produced by falling, rolling, or sliding.

tectonic

Relating to the forces and the movements of Earth and its lithosphere. Earthquakes, volcanoes, and mountain building are related to tectonic activity.

terrestrial

Relating to land or the planet Earth. Refers to the land above sea level.

thermocline

A thin region of rapid temperature change separating the warm waters of the upper ocean from the cold waters of the abyssal ocean.

thermohaline circulation

Circulation or movement of ocean water masses resulting from density differences caused by variation of temperature and salinity.

tidal basin

A body of water, natural or manmade, subject to tidal action.

tilted laminations

Thin layers of rock or sediment oriented at an angle from the horizontal.

topography

Surface relief of the land. Topography is usually measured in meters above sea level.

trade winds

Part of one of the three major circulation cells in each hemisphere, the trade winds exist from approximately 0° to 30° north or south latitude. Within the regions of the trade winds, prevailing winds blow easterly, or toward the west.

transition zone

A region within the mantle that separates the upper and the lower mantle that is characterized by a rapid increase in seismic wave velocities. The depth of the transition zone varies from 410 to 1,000 kilometers beneath the Earth's surface.

tributary

A stream that flows into or joins a larger stream or another body of water.

troposphere

The lowest layer of Earth's atmosphere in contact with Earth's surface. Most weather occurs within the troposphere. The troposphere extends from Earth's surface to elevations of approximately 10–15 kilometers.

tsunami

Enormous ocean wave produced by an underwater earthquake, landslide or volcanic eruption.

ultraviolet radiation (UV)

A portion of the electromagnetic spectrum. Ultraviolet radiation has a shorter wavelength than visible light.

unconformity

An erosional surface representing a gap in the geologic record between rock layers of different ages indicating that deposition was not continuous.

Uniformitarianism

The principle that the same geological processes and natural laws that modify the Earth's crust operating today have operated in the same manner or way throughout geologic time.

upper mantle

The part of the mantle which lies above the transition zone. It is presumed to be composed of peridotite.

upwelling

In the solid earth, the act of warm mantle rising upwards relative to surrounding, cooler mantle. In the ocean, the rising of deeper waters to replace migrating surface waters. Upwelling may bring waters rich in nutrients to the surface, resulting in a region where ocean productivity is high.

viscosity

The property of a substance which determines the amount of its internal resistance to flow. The opposite of fluidity.

visible light

The portion of the electromagnetic spectrum with wavelengths shorter than infrared radiation but longer than ultraviolet radiation. Visible light includes wavelengths that include all the colors we see: violet, blue, green, yellow, orange, and red light from the Sun.

volatiles

Elements and compounds that evaporate readily under normal surface conditions.

volcanism

Any of the processes by which magma and its associated gases rise to the crust from the Earth's interior and are discharged onto the surface and into the atmosphere.

volcano

A vent or fissure in the Earth's surface through which molten lava, ash, and gases are ejected. It is also the name for the structure, usually conical, formed by the materials ejected from the vent or fissure.

volcanology

The study of the causes and phenomena associated with volcanoes and volcanism.

X-ray

Electromagnetic radiation of short wavelength. The wavelengths of X-rays are shorter than those of ultraviolet radiation (which are of shorter wavelength than visible light).

zircon

A durable mineral ($ZrSiO_4$) that occurs in tetragonal prisms, has various colors, and is found in many different rock types.

Questions

Section One

An Earth Moon Mystery

In view of Dr. Noom's new data, which theory do you think best explains the Moon's formation?

How do the four models of lunar formation differ?

How does the analysis of isotopic chemical composition of the Earth and the Moon give us clues as to how the Moon may have formed?

Origin and Evolution of the Continents

How do trace elements help scientists determine the origin of continental crust?

What are some of the problems associated with the "andesite model"?

Life and the Evolution of Earth's Atmosphere

What factors have affected the evolution of Earth's atmosphere?

How have astrophysical studies of young stars in the galaxy affected ideas of Earth's early atmospheric conditions?

What is the relationship between the biosphere and the atmosphere?

Section Two

Global Seismic Tomography: A Snapshot of Convection in the Earth

Why have geophysicists sought to understand the pattern of convection within the mantle?

How are seismic waves used to determine the temperature and composition of Earth's interior?

How do scientists study convective flow in the mantle?

Convection in the Core and the Generation of the Earth's Magnetic Field

What can scientists learn from a computer model of Earth's core?

What could happen to living organisms on Earth if the geomagnetic field weakened?

Mantle Convection

What is the relationship between mantle convection and plate tectonics?

How does the use of computer modeling help geophysicists understand the dynamic evolution of the Earth?

Section Three

The Structure of Mountain Ranges

What are the underlying structures associated with various mountain ranges?

Why is the thickness of Earth's crust important to the mountain building process?

How do the structures of the Himalaya, the Tibetan plateau, the Alps, the Rocky Mountains, and the Andes differ? Which have similar structures?

Why are the Andes and Tibet particularly susceptible to collapse (in a geologic time scale)?

Averting Earthquake Surprises in the Pacific Northwest

How are seismologists able to detect the occurrence of ancient earthquakes?

Why is it important to design a Uniform Building Code for areas that experience natural hazards?

Hawaii and Hotspots: A Window to the Deep Mantle

How are scientists able to determine the direction of plate movement and age of islands that formed over hotspot regions?

Hotspots are located in oceanic crust as well as continental crust. How can they be recognized and what effect do they have on the surface environment?

What is the connection between hotspots and plate tectonics?

The Hazards of Volcanoes

How does magma change as it travels from the magma chamber to the crater opening?

How does the water content of magma affect the nature of a volcanic eruption?

Section Four

The Erosion of the Grand Canyon

What are some of the reasons why geologists disagree about how the Grand Canyon formed?

What geologic factors eroded and sculpted the Grand Canyon?

Death Among the Dunes: A Dinosaur Murder Mystery

What are some of the possible reasons that the organisms at Ukhaa Tolgod died?

What correlations can paleontologists derive between fossil deposits and the climate of past environments?

Section Five

The Oceans' Role in Climate

What is the relationship between ocean currents and global climate?

How do El Niño and La Niña affect the way oceans store and transport heat?

Why would it be valuable to predict El Niño?

Predicting El Niño

What is the relationship between El Niño and the Southern Oscillation?

Why is historical data important in finding patterns in climate?

Global Warming

Why is it so difficult for climatologists to predict global warming?

What are the natural causes of global warming? What are the human causes?

Why is it important to compare our current climate to past climate?

Why do we use computer models to study global warming?

What are the potential impacts of global warming?

An Ice Core Time Machine

Why are ice cores used to study the chemistry of Earth's past atmosphere?

What other natural records can be used to document past climate?

What kinds of climate events are recorded in the ice core?

Section Six

Earth: The Goldilocks Planet

What would make Earth more like Mars? More like Venus?

Where is Earth's carbon dioxide stored?

How is it transported between the different reservoirs?

Black Smokers: Incubators on the Seafloor

What are some of the reasons why black smoker chimney systems vary in shape at different spreading environments?

What implications do the discovery of black smokers have in the search for life on other planets?

Resources of the Earth: Where Do Metals Come From?

What are some factors mining companies consider before excavating metal-rich rocks?

How are hydrothermal fluids instrumental in the formation of metal deposits deep within the crust?